ANALYSIS FOR ENGINEERS

CHARLES W. HAINES
Rochester Institute of Technology

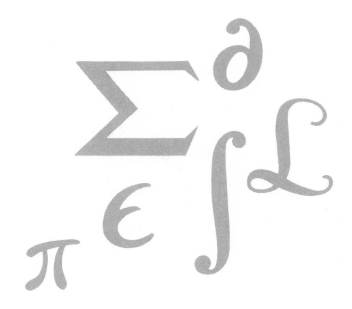

WEST PUBLISHING CO.
St. Paul • New York • Boston
Los Angeles • San Franciso

Library of Congress Catalog Card Number: 74–866
ISBN: 0–8299–0011–X

Haines—Analysis for Engineers

In Memory Of
Clifford C. Crump
Astronomer, Teacher and Friend

*

PREFACE

This book is written for engineering and science students who have had two or
three semesters of differential and integral calculus. Students in such other
diverse areas as the biological sciences and economics will increasingly need some
understanding of the basic applied mathematics presented here. The material is also
appropriate for those in the engineering, scientific, technical and teaching pro-
fessions who need to refresh or upgrade their mathematical knowledge.

Each of the chapters is designed to present the highlights of the subject
matter that are most appropriate for applications without covering all possible
details and extensions. In this way a broad coverage of the mathematics at the in-
termediate level has been attained.

The presentation of the material is meant to be clear and accurate but intuitive
rather than rigorous. It is definitely meant for engineers and students in other
scientific areas, and thus the topics, examples and exercises have been chosen with
this in mind. Theory is discussed where appropriate for a good understanding of the
applicability, while various pitfalls have been noted to indicate the limits of
usefulness of the material. Many examples and figures have been presented to supple-
ment and clarify the text while a large number of exercises allow the student to
work on problems ranging from the routine to the development of additional results.

The book is probably not suitable for a traditional semester or quarter course
in any one of the chapter areas since the presentation does not go deeply enough
into such topics as complex variables or differential equations. However, it is
felt that much of the material needed by engineers or scientists at this level is
presented so that it is appropriate to use the text for a fourth or fifth semester
of mathematics in the engineering curriculum. For instance, a course based on
Chapters 5 and 6 (Vector and Complex Analysis) could very well follow the traditional
introductory calculus and differential equations sequence, a course based on Chapters

1, 2, 7 and 8 gives an introduction to various topics found in linear analysis, or a course based on Chapters 1-5 (or Chapters 1-5 and 7) could very well be used to cover the essential material of differential equations and differential and integral calculus. The material has been written so that the chapters are more or less independent of each other and hence there is an even wider selection of topics that is possible for a particular course than those indicated above. The Laplace transform chapter is more comprehensive if preceded by complex variables but, except for the work on the inverse transform in Section 7.5, can be studied without Chapter 6. Examples and exercises in Chapter 8 rely heavily on a knowledge of differential equations (Chapter 1) while the last three sections of Chapter 5 require the material from Chapter 4 on integration.

The broad coverage of material is also quite suitable for continuing education courses. For instance, at Rochester Institute of Technology most of Chapters 1 through 5 have been used for a refresher course for students in Mechanical and Industrial Engineering while Chapters 1-4 and 6 have been used for a similar course for Electrical Engineering students. Both of these courses have been taught successfully at several local industrial locations using a video tape format. Each of the chapter sections is presented as a video tape unit and is viewed at scheduled or irregular times by the students. Once a week the author conducted a live recitation for the purpose of answering questions on the text, reviewing assigned exercises and developing topics further where appropriate. Using the modularized format of the book, many combinations of courses can be taught for the continuing education student.

I wish to express my appreciation to Professor David Powers of Clarkson College of Technology, who read and offered critical commentary on the entire manuscript; Professors Watson Walker, Robert Desmond, Richard Kenyon and John Paliouras of Rochester Institute of Technology, who contributed through many discussions; Ms. Ann Towne for help in proofreading and indexing; Mr. Francis Janucik, Mr. Charles Thomas and Mr. Alfred Wicks for help in checking answers; Mrs. Patricia Szulc, who very quickly and efficiently typed the manuscript; and my wife Carolyn for her encouragement and understanding throughout the whole process.

Rochester, N. Y.   *Charles W. Haines*
January 1974

CONTENTS

1

DIFFERENTIAL EQUATIONS

1.1 INTRODUCTION AND FIRST ORDER EQUATIONS

Many significant problems in engineering and other disciplines, when formulated
in mathematical terms, require the determination of a function satisfying an equation
containing derivatives of the unknown function. Such equations are called <u>differential
equations</u>. We will mainly be concerned with one particular type: linear ordinary
differential equations. <u>Ordinary</u> means the unknown function depends on only one
variable (such as time) and <u>linear</u> means that the unknown function and its derivatives
are raised only to the first power; in addition terms like $y(t) \frac{dy}{dt}$ and $\sin y(t)$ are
also excluded from linear equations since they are not linear functions of the unknown variable $y(t)$. Some examples from engineering problems are:

Example 1.   The charge $Q(t)$ on the capacitor in the electrical circuit
    shown in Figure 1. is determined by the equation

$$R \frac{dQ}{dt} \;+\; \frac{1}{C} Q(t) = E(t), \tag{1}$$

where R is the resistance, C is the capacitance and $E(t)$ is the
impressed voltage.

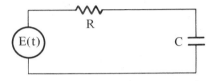

Figure 1.

Example 2.   The displacement $x(t)$ of the mass in the mechanical
    spring system of Figure 2. is determined by the equation

$$m \frac{d^2x}{dt^2} + b \frac{dx}{dt} + k\,x(t) = f(t), \tag{2}$$

where m is the mass, b the damping constant, and k the spring constant.

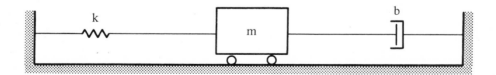

Figure 2.

Examples from electrical, chemical and industrial engineering also give rise to equations similar to Eq. (2). Examples 1. and 2. are both <u>constant</u> coefficient equations since it is assumed R, C, m, b and k are constant. Example 1. is a <u>first</u> <u>order</u> equation and Example 2. is a <u>second</u> <u>order</u> equation since these are the highest derivatives appearing in the differential equations. Both Eqs. (1) and (2) are <u>non-homogeneous</u>. If f(t) and E(t) were both zero, then the equations would be <u>homogeneous</u>. Homogeneous here essentially means all terms involve the same variable such as x(t) in Eq. (2) or Q(t) in Eq. (1) when the right hand sides are set equal to zero. Finally, in order to find a unique solution to either Eq. (1) or (2) appropriate <u>initial</u> <u>conditions</u> need to be specified, such as x(0) and $\frac{dx}{dt}$ (0) in Example 3.

Example 3.

$$2 \frac{d^2x}{dt^2} + \frac{dx}{dt} + x = 2 \sin t,$$

$$x(0) = 1, \qquad \frac{dx}{dt}(0) = 0.$$

A first order equation would require only one initial condition rather than the two shown for the second order equation of Example 3.

The technique of solving linear constant coefficient equations will be demonstrated by solving and studying in detail first and second order equations. Relations with the physical world will be used throughout to maintain the application-oriented approach.

The general first order linear differential equation is:

$$\frac{dy}{dt} + \alpha y(t) = f(t). \tag{3}$$

The <u>general</u> solution to Eq. (3) is

$$y_g = y_h(t) + y_p(t) \tag{4}$$

where y_h is the solution to the related homogeneous equation:

$$\frac{dy_h}{dt} + \alpha y_h(t) = 0, \tag{5}$$

and $y_p(t)$ is <u>any</u> solution of Eq. (3). This fact can be established by differentiating y_g, as given by Eq. (4), and then adding that to αy_g as follows:

$$\frac{dy_g}{dt} + \alpha y_g = \frac{d}{dt}(y_h + y_p) + \alpha(y_h + y_p) \tag{6}$$

$$= \frac{dy_h}{dt} + \frac{dy_p}{dt} + \alpha y_h + \alpha y_p \tag{7}$$

$$= \frac{dy_h}{dt} + \alpha y_h + \frac{dy_p}{dt} + \alpha y_p \tag{8}$$

$$= 0 + f(t), \tag{9}$$

where Eq. (5) and Eq. (3) have been employed.

Equation (4), which is valid only for linear equations, is a very important result that will be seen again in Section 1.2 and later in the solution of linear systems of equations. By using Eq. (4), we may solve a linear non-homogeneous equation in two parts: the related homogeneous solution y_h is found using one method and then the particular solution y_p is found using a completely different approach. These will be shown in the rest of this section and the next.

To find y_h, which is also called the complementary solution, we assume a solution of the form

$$y_h = ce^{rt}. \tag{10}$$

The basis for this assumption can be seen by rewriting Eq. (5) as

$$\frac{dy_h}{dt} = -\alpha\, y_h,$$

and hence what we are seeking is a function whose derivative is a constant multiple of the function itself. From elementary calculus the only such function is the exponential function assumed in Eq. (10). By differentiating Eq. (10) we obtain

$$\frac{dy_h}{dt} = r\, ce^{rt}$$

and thus

$$\frac{dy_h}{dt} + \alpha y_h = r\, ce^{rt} + \alpha\, ce^{rt}$$

$$= (r+\alpha)\, ce^{rt}. \tag{11}$$

For y_h to satisfy Eq. (5), the right hand side of Eq. (11) must be zero. Since the exponential function is never zero and $c = 0$ would yield the trivial solution ($y_h(t) = 0$ for all t) we must conclude that

$$r = -\alpha .$$

Thus

$$y_h = ce^{-\alpha t} \tag{12}$$

satisfies Eq. (5) for any constant c. It should be noted that the complementary solution involves the arbitrary constant c. At this point the choice of c is immaterial, but when initial conditions need to be satisfied, then we must choose c to satisfy them.

To find the particular solution y_p is a more complicated process and depends upon the function f(t) in Eq. (3). In most engineering problems f(t) is a polynomial, a sinusoid or an exponential, in which case $y_p(t)$ is assumed to be of a similar form with undetermined coefficients. Then to find y_p, simply substitute the assumed form into Eq. (3) and solve for the undetermined coefficients.

The appropriate form to assume is obtained by adding f(t) and all functions obtained by differentiating f(t), each multiplied by an undetermined coefficient. For instance, if

$$f(t) = \cos 2t \tag{13}$$

then y_p has the assumed form:

$$y_p(t) = A \sin 2t + B \cos 2t. \tag{14}$$

The following examples and exercises will further illustrate this procedure for finding the particular solution.

Example 4.   Find the general solution of the differential equation

$$\frac{dy}{dt} + 3y = e^{-t}. \tag{15}$$

From Eq. (12) we see that $y_h = ce^{-3t}$ and from above we assume that y_p has the form

$$y_p = Ae^{-t}.$$

Then

$$\frac{dy_p}{dt} + 3y_p = -Ae^{-t} + 3Ae^{-t}$$

$$= 2A e^{-t}.$$

Thus y_p will be a solution of Eq. (15) provided $A = \frac{1}{2}$.

Hence

$$y_g = ce^{-3t} + \tfrac{1}{2}e^{-t} \tag{16}$$

is the general solution to Eq. (15).

Example 5.   Solve Eq. (1) when $E(t)$ is a constant voltage E_0
                      and there is no initial charge on the capacitor.

The mathematical problem is then:

$$R\frac{dQ}{dt} + \frac{1}{C}Q = E_0 \tag{17}$$

$$Q(0) = 0. \tag{18}$$

The solution of the related homogeneous equation is

$$Q_h = ce^{-t/RC} \tag{19}$$

and the assumed form for y_p is

$$Q_p = A. \tag{20}$$

Substituting this into Eq. (17) we get:

$$R\frac{dQ_p}{dt} + \frac{1}{C}Q_p = 0 + \frac{1}{C}A.$$

Thus Q_p will be a solution of Eq. (17) provided $A = CE_0$.
Substituting this value for A in Eq. (20), we arrive at

$$Q(t) = ce^{-t/RC} + CE_0 \tag{21}$$

as the general solution of Eq. (17). To satisfy the initial
condition (18) we let $t = 0$ in Eq. (21):

$$Q(0) = c + CE_0. \tag{22}$$

Consequently we want $c = -CE_0$, and conclude that

$$Q(t) = CE_0 - CE_0e^{-t/RC}$$

or

$$Q(t) = CE_0(1 - e^{-t/RC}). \tag{23}$$

An analysis of Eq. (23) shows that at $t = 0$ there is indeed zero charge on the
capacitor (the initial condition); and as time increases the charge increases,
approaching the final value of CE_0. A sketch of the solution is shown in Figure 3.

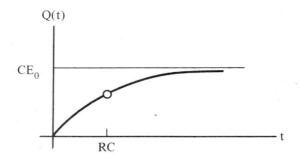

Figure 3.

An alternate method of solving Eq. (3) is to use an <u>integrating factor</u>. The advantage of this approach is that an integrating factor can be found even when α is a variable coefficient:

$$\frac{dy}{dt} + \alpha(t)y = f(t) \ . \tag{24}$$

The integrating factor is a function $\mu(t)$ which multiplies the left hand side of Eq. (24) in such a way that an exact derivative is obtained. That is:

$$\mu(t)(\frac{dy}{dt} + \alpha(t)y) = \frac{d}{dt}(\mu y). \tag{25}$$

Once $\mu(t)$ is found from Eq. (25), then Eq. (24) can be solved by multiplying both sides by $\mu(t)$ and using Eq. (25) for the left side:

$$\mu(t)(\frac{dy}{dt} + \alpha(t)y) = \mu(t) \ f(t)$$

or

$$\frac{d}{dt}(\mu y) = \mu(t) \ f(t). \tag{26}$$

This last equation can be integrated since it shows that $\mu(t) \ f(t)$ is the derivative of $\mu(t) \ y(t)$. Thus $y(t)$ can be obtained by dividing by $\mu(t)$:

$$y(t) = \frac{1}{\mu(t)} \int \mu(t) \ f(t) \ dt + \frac{c}{\mu(t)} \ . \tag{27}$$

Once $\mu(t)$ is known, then Eq. (27) gives the solution to Eq. (24). An example will illustrate its use after we see how to find $\mu(t)$.

To find $\mu(t)$ expand the right hand side of Eq. (25):

$$\frac{d}{dt}(\mu y) = \frac{d\mu}{dt} y + \mu \frac{dy}{dt}$$

and equate to the left hand side:

$$\frac{d\mu}{dt} y + \mu \frac{dy}{dt} = \mu \frac{dy}{dt} + \mu\alpha y$$

or

$$\frac{d\mu}{dt} y = \mu\alpha y . \tag{28}$$

While Eq. (28) does not appear any simpler than the original Eq. (24), we should note that it is homogeneous in $\mu(t)$ and by division can be put in the form:

$$\frac{1}{\mu} \frac{d\mu}{dt} = \alpha(t), \tag{29}$$

assuming $\mu \neq 0$. The left hand side of Eq. (29) is recognized as the derivative of $\ln \mu(t)$ and hence integration yields

$$\ln \mu(t) = \int \alpha(t) dt, \tag{30}$$

or

$$\mu(t) = e^{\int \alpha(t)dt} . \tag{31}$$

In theory, then, Eq. (31) substituted into Eq. (27) will yield the solution to the linear equation with a variable coefficient. In practice the integrals, even though they exist mathematically, may not be elementary functions and hence Eq. (27) will yield only an integral representation of the general solution. The following example will illustrate this method.

Example 6.   Find the general solution of

$$\frac{dy}{dt} + \frac{1}{t} y = t, \qquad t > 0. \tag{32}$$

We must first of all find $\mu(t)$. From Eq. (32) $\alpha(t)$ is found to be $\frac{1}{t}$ and hence Eq. (30) yields:

$$\ln \mu(t) = \int \frac{dt}{t}$$

$$= \ln t.$$

Thus $\mu(t) = t$ and Eq. (27) becomes:

$$y(t) = \frac{1}{t} \int t \cdot t \, dt + \frac{c}{t} \tag{33}$$

since $f(t) = t$ also. The general solution $y(t)$ is easily found
from Eq. (33) as

$$y(t) = \frac{t^2}{3} + \frac{c}{t} .$$

(34)

Exercises

1. Use undetermined coefficients to find the general solution of each of the
    following problems.

    a) $\frac{dy}{dx} + y = 1 + 2x$

    b) $\frac{dy}{dx} + 3y = x + e^{-2x}$

    c) $\frac{dy}{dt} - y = \sin t$

    d) $\frac{dy}{dx} + 2y = xe^{-x}$

2. Solve each of the following initial value problems and sketch the solution.

    a) $\frac{dy}{dt} + y = t,$ $y(0) = 0$

    b) $3\frac{dy}{dt} + y = 2,$ $y(0) = 1$

    c) $L\frac{di}{dt} + Ri = E_0 \sin 2t,$ $i(0) = 0$

       (Set $L = 1$, $R = 2$ for the sketch)

3. Show that $y_p = Ae^{-t}$ is not an appropriate form for the particular solution of

    $$\frac{dy}{dt} + y = 2e^{-t}.$$

    Explain why, and show that $y_p = Ate^{-t}$ does yield a particular solution.

4. Find the general solution of each of the following problems.

    a) $\frac{dy}{dx} + \frac{2}{x}y = \frac{\sin x}{x^2}$

    b) $t\frac{dy}{dt} + y = \cos t$ (Hint: $\alpha(t) \neq 1$.)

5. Solve the initial value problem

    $$\frac{dy}{dt} - 2ty = 2t,   y(0) = 2.$$

6. Show that Eq. (27) becomes

$$y(t) = e^{-\alpha t} \int e^{\alpha t} f(t) \, dt + c e^{-\alpha t}$$

when $\alpha(t)$ is a constant. This gives a general form of the solution of Eq. (3) when undetermined coefficients do not apply.

7. Use the results of Exercise 6. to find the general solution of

$$\frac{dy}{dt} + y = t + e^{-2t}.$$

1.2   SOLUTION OF SECOND ORDER EQUATIONS

The solution of second order, linear, constant coefficient equations follows exactly the same pattern as was developed for the first order constant coefficient equations. The general equation of this type is

$$\frac{d^2 y}{dt^2} + a \frac{dy}{dt} + b \, y(t) = f(t). \tag{1}$$

Again, the general solution of Eq. (1) is found in the form

$$y_g(t) = y_h(t) + y_p(t) \tag{2}$$

where $y_h(t)$ satisfies the related homogeneous equation

$$y_h'' + a y_h' + b y_h = 0 \tag{3}$$

and $y_p(t)$ is <u>any</u> solution of Eq. (1). In Eq. (3) and throughout the rest of this chapter the primes are used to denote the derivatives.

That $y_g(t)$ does satisfy Eq. (1) follows from the fact that Eq. (1) is linear and the proof would follow the same steps as done in Section 1.1 for the first order case.

The procedure for finding $y_p(t)$ is exactly the same as was developed in Section 1.1 for the solution of first order equations, since again the non-homogeneous term $f(t)$ for most engineering problems will be a polynomial, sinusoid, or an exponential.

To solve the homogeneous Eq. (3) we again assume that the solution has the form

$$y_h(t) = c e^{rt} \tag{4}$$

which, when substituted into Eq. (3), yields

$$y_h'' + a \, y_h' + b \, y_h = r^2 c e^{rt} + a \, r \, c e^{rt} + b \, c e^{rt}$$

$$= (r^2 + ar + b) c e^{rt}. \tag{5}$$

The right side of Eq. (5) must be zero if ce^{rt} is to satisfy Eq. (3). Since c = 0 would yield the trivial solution and e^{rt} is never zero, we then must conclude that

$$r^2 + ar + b = 0. \tag{6}$$

This is called the <u>characteristic equation</u> of Eq. (3), since the solutions of Eq. (6) determine the nature of the solutions of Eq. (3). Equation (6) is a quadratic so that there will be <u>two</u> values of r, say r_1 and r_2, which satisfy it. Thus we have found <u>two solutions</u>, $y_1(t) = c_1 e^{r_1 t}$ and $y_2(t) = c_2 e^{r_2 t}$, which satisfy Eq. (3). The sum of $y_1(t)$ and $y_2(t)$ will also satisfy Eq. (3) since it is linear. Thus the general solution of Eq. (3) is

$$y_h(t) = y_1(t) + y_2(t) \tag{7}$$

$$= c_1 e^{r_1 t} + c_2 e^{r_2 t}. \tag{8}$$

The student should verify, by differentiating Eq. (8) and substituting into Eq. (3), that it indeed satisfies Eq. (3), provided r_1 and r_2 satisfy Eq. (6).

Equation (7), which involves two arbitrary constants, is known as the <u>general solution</u> of the homogeneous Eq. (3). It is assumed in Eq. (7) that $y_1(t)$ and $y_2(t)$ are <u>linearly independent</u> solutions of Eq. (3). Linear independence is a mathematical term which here means that $y_1(t)$ and $y_2(t)$ <u>are</u> <u>not</u> constant multiples of each other. If $y_2(t)$ were a multiple of $y_1(t)$, then Eq. (7) would not involve <u>two</u> arbitrary constants. Later examples will illustrate this point.

One thing that we have not considered yet is the nature of the roots r_1 and r_2. Since Eq. (6) is a quadratic with real coefficients, the roots r_1 and r_2 can be either real and unequal, real and equal, or complex conjugates. For the last two cases the form of the solution of Eq. (3) is different. Rather than develop these cases, we shall consider several examples which will illustrate the different situations.

Example 1.   Find the general solution of

$$y'' + 4y' + 3y = 0. \tag{9}$$

If we assume $y(t) = ce^{rt}$, then

$$y'' + 4y' + 3y = r^2 ce^{rt} + 4rce^{rt} + 3ce^{rt}$$

$$= (r^2 + 4r + 3)ce^{rt}. \tag{10}$$

If the right side is to be zero for all t, we must have

$$r^2 + 4r + 3 = 0. \tag{11}$$

The roots of Eq. (11) are $r = -1$ and $r = -3$. Therefore

$$y(t) = c_1 e^{-t} + c_2 e^{-3t} \tag{12}$$

is the general solution of Eq. (9).

Example 2.   Find the general solution of

$$y'' + 2y' + y = 0. \tag{13}$$

Again let $y(t) = ce^{rt}$, then as before

$$y'' + 2y' + y = (r^2 + 2r + 1) ce^{rt}$$

which implies that

$$r^2 + 2r + 1 = 0$$

or

$$(r + 1)^2 = 0. \tag{14}$$

Thus the roots are real and equal. Unfortunately this leaves us with only one solution of Eq. (13): namely, $y_1(t) = c_1 e^{-t}$, since the second solution would be $y_2(t) = c_2 e^{-t}$ if the previous pattern were followed. Since c_2 is arbitrary, this is exactly the same as $y_1(t)$ and thus the solutions are not linearly independent.

The second linearly independent solution (if the roots are repeated) is obtained by multiplying the first solution by t. For Example 2. then, the second linearly independent solution is $y_2(t) = c_2 te^{-t}$. Thus

$$y(t) = c_1 e^{-t} + c_2 te^{-t} \tag{15}$$

is the general solution of Eq. (13). The student should verify for Example 2. that te^{-t} is indeed a solution.

Example 3.   Find the general solution of

$$y'' + \omega^2 y = 0. \tag{16}$$

Again, if $y(t) = ce^{rt}$, we find that r must satisfy

$$r^2 + \omega^2 = 0. \tag{17}$$

This yields two complex conjugate roots $\pm i\omega$, where $i = \sqrt{-1}$.
Hence the solution is of the form

$$y(t) = B_1 e^{i\omega t} + B_2 e^{-i\omega t}. \tag{18}$$

The solution of Eq. (16) as expressed in Eq. (18) is not satisfactory, since it is a complex valued answer to a real valued problem. By using the identity (known as Euler's formula)

$$e^{ix} = \cos x + i \sin x, \tag{19}$$

Eq. (18) can be rewritten in the form

$$y(t) = c_1 \cos \omega t + c_2 \sin \omega t . \tag{20}$$

Equation (20) represents the general solution to Eq. (16) and is applicable whenever the roots of the characteristic equation are complex conjugates.

Example 4.   Find the general solution of

$$y'' + 2y' + 2y = 0. \tag{21}$$

If $y = ce^{rt}$, we obtain the characteristic equation

$$r^2 + 2r + 2 = 0. \tag{22}$$

The roots of Eq. (22) are $r = -1 \pm i$. Thus the solution of Eq. (21) will be

$$y(t) = B_1 e^{(-1 + i)t} + B_2 e^{(-1 - i)t} \tag{23}$$

$$= e^{-t}(B_1 e^{it} + B_2 e^{-it})$$

$$= e^{-t}(c_1 \cos t + c_2 \sin t), \tag{24}$$

which follows by using Euler's formula.

The above examples can be summarized by the following general results:

For the homogeneous equation

$$y'' + ay' + by = 0$$

with characteristic equation

$$r^2 + ar + b = 0$$

we have three cases:

1) real and unequal roots; r_1, r_2

$$y(t) = c_1 e^{r_1 t} + c_2 e^{r_2 t} \tag{25}$$

2) repeated roots; r_1, r_1

$$y(t) \; = \; (c_1 + c_2 t)e^{r_1 t} \tag{26}$$

3) complex conjugate roots; $\alpha \pm i \, \beta$

$$y(t) = e^{\alpha t}(c_1 \cos \beta t + c_2 \sin \beta t). \tag{27}$$

In each case $y(t)$ is the general solution of the homogeneous equation.

Exercises

1. Find the general solution of each of the following homogeneous equations.

a) $y'' + 4y' + 4y = 0$ b) $y'' + y' - 6y = 0$

c) $y'' - 2y' + 10y = 0$ d) $y'' + y' + y = 0$

2. Find the general solution of each of the following non-homogeneous equations.

a) $y'' + 4y = 1$

b) $y'' + 6y' + 9y = e^{-t}$

c) $y'' + y = \sin 2t$

3. Find the unique solution of each of the following initial value problems.

a) $y'' + 5y' + 6y = 1$, $y(0) = 1$, $y'(0) = 0$

b) $y'' + y' - 2y = 2t$, $y(0) = 0$, $y'(0) = 1$

c) $y'' + 3y = \cos t$, $y(0) = 0$, $y'(0) = 1$

4. Use the identity

$$\sin(\beta t + \phi) = \sin \beta t \cos \phi + \cos \beta t \sin \phi$$

to show that the solution to Example 3. can be written as

$$y(t) = R \sin(\omega t + \phi).$$

Find R and ϕ in terms of c_1 and c_2 of Eq. (20).

1.3 APPLICATIONS OF SECOND ORDER EQUATIONS AND n^{th} ORDER EQUATIONS

The previous section and exercises were concerned with the solution of the equation

$$\frac{d^2y}{dt^2} + a\frac{dy}{dt} + b\, y(t) = f(t) \, . \tag{1}$$

In this section we shall be considering the general solution of Eq. (1) and relating its behavior to some physical examples. You will recall in Section 1.1 that a mechanical mass, spring, damper system gave rise to the differential equation

$$m\frac{d^2x}{dt^2} + b\frac{dx}{dt} + k\, x = f(t) \tag{2}$$

where m is the mass, b is the damping coefficient, k is the spring constant and f(t) is an external forcing function. Similar equations also arise in other applications. For instance, an electric circuit with inductance L, capacitance C, resistance R and impressed voltage E(t) will yield

$$L\frac{d^2Q}{dt^2} + R\frac{dQ}{dt} + \frac{1}{C}Q = E(t) \tag{3}$$

as the equation governing the charge Q(t) on the capacitor. In both Eqs. (2) and (3) the constant coefficients represent physical quantities and therefore may be assumed to be positive.

In this section we deal exclusively with Eq. (2), but the analogies with Eq. (3) will be straightforward, if that equation is more familiar to the student.

We first of all divide Eq. (2) by m to obtain

$$\frac{d^2x}{dt^2} + 2\alpha\frac{dx}{dt} + \omega^2 x = g(t) \tag{4}$$

where

$$2\alpha = \frac{b}{m}, \quad \omega^2 = \frac{k}{m} \quad \text{and} \quad g(t) = \frac{f(t)}{m} \, .$$

Here ω is called the "natural frequency" since this is the frequency of vibration when there is no damping. (See Example 3. of Section 1.2.)

In the last section we saw that the general solution of Eq. (4) would be written in the form

$$x(t) = x_h(t) + x_p(t), \tag{5}$$

where x_h satisfies the homogeneous equation

$$x'' + 2\alpha x' + \omega^2 x = 0, \tag{6}$$

and $x_p(t)$ is any solution of Eq. (4). The first important physical result is that if $b \neq 0$ -- there is damping in the system -- then $x_h(t)$ approaches zero as t becomes large. We can show this by looking at the roots of the characteristic equation. For Eq. (6) we have the characteristic equation

$$r^2 + 2\alpha r + \omega^2 = 0, \tag{7}$$

which has the roots

$$r = -\alpha \pm \sqrt{\alpha^2 - \omega^2}. \tag{8}$$

Thus, since $\alpha > 0$, both values of r will be negative when r is real. When r is complex $(\alpha^2 - \omega^2 < 0)$, the real part of the root is negative. Thus, for all three cases the solution will have a negative-exponential time dependence (see Eqs. 25, 26, 27 in Section 1.2) and hence decay to zero as t gets large. For this reason $x_h(t)$ is also commonly called the transient solution.

Returning to Eq. (5), if $x_h(t)$ approaches zero as t gets large, then $x_p(t)$ will be the remaining part of the solution. Thus $x_p(t)$ is also commonly called the steady-state solution. The following example will illustrate the above concepts and will illustrate the behavior of physical systems.

Example 1.   Study the solution of the problem

$$x'' + 2\alpha x' + 4x = 4, \tag{9}$$

$$x(0) = 0, \quad x'(0) = 0, \tag{10}$$

for various α.

The particular solution is $x_p(t) = 1$, which can be checked easily. The behavior of the solution of Eq. (9) will be given by the form of the general solution $x_h(t)$ of the related homogeneous equation

$$x'' + 2\alpha x' + 4x = 0, \tag{11}$$

which has the characteristic equation

$$r^2 + 2\alpha r + 4 = 0. \tag{12}$$

The roots of Eq. (12) are

$$r = -\alpha \pm \sqrt{\alpha^2 - 4} \tag{13}$$

which clearly depend upon the value of α. If the system has "light" damping, $\alpha^2 < 4$, the roots are complex conjugates and $x_h(t)$ will be given by

$$x_h(t) = e^{-\alpha t}(c_1\cos \beta t + c_2\sin \beta t),$$

where

$$\beta = \sqrt{4 - \alpha^2} \ .$$

Thus, the general solution of Eq. (9) is:

$$x(t) = 1 + e^{-\alpha t}(c_1\cos \beta t + c_2\sin \beta t). \tag{14}$$

The initial conditions (10) require that

$$c_1 = -1 \text{ and } c_2 = \frac{-\alpha}{\beta} \ .$$

Thus

$$x(t) = 1 - e^{-\alpha t}(\cos \beta t + \frac{\alpha}{\beta} \sin \beta t). \tag{15}$$

Notice that as t gets large the effect of the last two terms is less and $x(t)$ gets closer to its steady-state value. This is shown in Figure 1. for very low values of damping. For larger values of

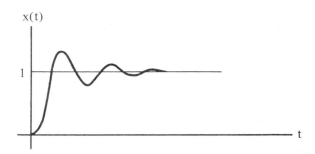

Figure 1.

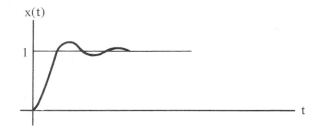

Figure 2.

damping the graph would vary slightly, as seen in Figure 2, for $\alpha < 2$ but close to 2. For a value of $\alpha > 2$, the form of the solution is

$$x(t) = 1 + c_1 e^{r_1 t} + c_2 e^{r_2 t} \tag{16}$$

since the roots of the characteristic equation, Eq. (12), are now real and unequal. The constants c_1 and c_2 can again be found from the initial conditions (10), but the important feature here is that the oscillatory nature of the previous solutions has disappeared, as illustrated in Figure 3.

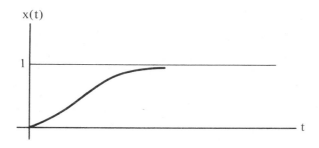

Figure 3.

The solution to Example 1. when $\alpha < 2$ can also be written as

$$x(t) = 1 - Re^{-\alpha t} \sin(\beta t + \phi), \tag{17}$$

where $R \cos \phi = \alpha/\beta$ and $R \sin \phi = 1$. This follows from Eq. (15) directly by the use of the identity

$$\sin(\beta t + \phi) = \sin \beta t \cos \phi + \cos \beta t \sin \phi. \tag{18}$$

The form of the solution as given in Eq. (17) is completely general and fully equivalent to the form shown in Eq. (27) of Section 1.2 for the case of complex roots (also see Exercise 4. of the last section).

A second application of differential equations introduces a different type of problem than we have encountered so far. Up to this point, in order to find values for our two arbitrary constants in the general solution of a differential equation we have specified two initial conditions. The initial conditions specify the function and its slope at an initial time (usually zero). In the application considered here, __boundary__ __conditions__ are specified rather than initial conditions. An example of this type of problem is given in Example 2.

Example 2.
$$\frac{d^2 y}{dx^2} + \omega^2 y = 1, \qquad \omega^2 > 0 \tag{19}$$

$$y(0) = 1, \qquad y(1) = 0. \tag{20}$$

As we see in Example 2, boundary conditions specify the function (or its slope) at two different points -- the boundary points. A detailed study of these problems will not be presented, but the procedure would be to find the general solution of the differential equation and then <u>try</u> to satisfy the boundary conditions. We say <u>try</u>, since boundary condition problems may have no solution, a unique solution or an infinite number of solutions, depending on the boundary conditions and the non-homogeneous term in the differential equation. This procedure is illustrated in Example 3. where the solution of Example 2. is discussed.

Example 3.   Find the solution of the boundary value problem of
                    Example 2.

The general solution of the differential equation is

$$y(x) = \frac{1}{\omega^2} + c_1 \cos \omega x + c_2 \sin \omega x, \tag{21}$$

as found by previously discussed methods. Setting $x = 0$ in Eq. (20) we obtain

$$y(0) = \frac{1}{\omega^2} + c_1$$

which must equal 1 if the first boundary condition is to be satisfied. Solving for c_1 we then have

$$y(x) = \frac{1}{\omega^2} + \frac{\omega^2 - 1}{\omega^2} \cos \omega x + c_2 \sin \omega x, \tag{22}$$

where c_2 must now be chosen to satisfy the second boundary condition, $y(1) = 0$. This will be satisfied provided

$$\frac{1}{\omega^2} + \frac{\omega^2 - 1}{\omega^2} \cos \omega + c_2 \sin \omega = 0, \tag{23}$$

which is obtained when $y(1)$ is found from Eq. (22). The difficulties with boundary conditions may now be seen as it is noted that Eq. (23) may not be solved for c_2 if $\omega = n\pi$, $n = 1,2,3...$ since then $\sin \omega = 0$ and c_2 does not appear in the equation. Thus the given problem has the unique solution

$$y(x) = \frac{1}{\omega^2} + \frac{\omega^2 - 1}{\omega^2} \cos \omega x + \frac{(1 - \omega^2)\cos \omega - 1}{\omega^2 \sin \omega} \sin \omega x,$$

provided $\omega \neq n\pi$, $n = 1,2,3\ldots$, and has no solution when $\omega = n\pi$, $n = 1,2,3\ldots$.

Finally, a few words should be mentioned concerning higher order differential equations. The general n^{th} order linear equation is:

$$a_n \frac{d^n y}{dt^n} + a_{n-1} \frac{d^{n-1} y}{dt^{n-1}} + \cdots + a_1 \frac{dy}{dt} + a_0 y = f(t). \tag{24}$$

The general solution of this equation is again of the form

$$y_g(t) = y_h(t) + y_p(t), \tag{25}$$

where $y_p(t)$ is any solution of Eq. (24) and $y_h(t)$ is the general solution of the related homogeneous equation

$$a_n \frac{d^n y}{dt^n} + a_{n-1} \frac{d^{n-1} y}{dt^{n-1}} + \cdots + a_1 \frac{dy}{dt} + a_0 y = 0. \tag{26}$$

The characteristic equation for Eq. (26) is

$$a_n r^n + a_{n-1} r^{n-1} + \cdots + a_1 r + a_0 = 0, \tag{27}$$

which has n roots, and hence $y_h(t)$ will involve n arbitrary constants:

$$y_h(t) = c_1 e^{r_1 t} + c_2 e^{r_2 t} + \cdots + c_n e^{r_n t}, \tag{28}$$

provided the n roots of the characteristic equation are distinct. The fact that the sum of solutions of the linear, homogeneous Eq. (26) (denoted as $c_i e^{r_i t}$) is also a solution of the equation is often referred to as the superposition principle. The same principle was also encountered in the previous section and is of fundamental importance in the solution of linear differential equations and linear systems of equations as studied in Chapter Two.

If the roots of the characteristic equation are not distinct then the appropriate terms in Eq. (28) must be multiplied by powers of t (the power depends on the multiplicity of the root) in order to make the n solutions linearly independent. If some of the roots are complex values, then the appropriate terms of Eq. (28) must be modified. Both of these changes have been described in more detail in the previous section and will not be discussed further here.

In summary, then, the pattern for solving higher order equations is exactly the same as the one we have used for second order equations. The main difference which does occur is in the calculation of the n roots of the characteristic equation, which usually would have to be done numerically.

Exercises

1. Solve the undamped vibration problem

$$x" + \omega^2 x = 1, \qquad x(0) = 1, \qquad x'(0) = 1.$$

Write the solution in the alternate form as shown by Eq. (17) and from this sketch the solution.

2. Solve the undamped forced vibration problem

$$x" + \omega^2 x = \cos 2t, \qquad x(0) = 0, \qquad x'(0) = 0.$$

What happens if $\omega^2 = 4$? Using the identities

$$\cos (A + B) = \cos A \cos B - \sin A \sin B$$
$$\cos (A - B) = \cos A \cos B + \sin A \sin B$$

rewrite the solution in the format

$$x(t) = c \sin \frac{\omega - 2}{2} t \sin \frac{\omega + 2}{2} t$$

and use this to sketch the solution when $\omega = 2.2$.

3. Determine the solution of

$$x" + 2x' + 2x = \sin 2t, \qquad x(0) = 0, \qquad x'(0) = 0.$$

What is the transient solution and what is the steady state solution? Again write the solution in the form of Eq. (17).

4. Solve (if possible) each of the following boundary value problems.

a) $\dfrac{d^2 y}{dx^2} + \lambda^2 y = 0, \qquad y(0) = 0, \qquad y(1) = 0$

(This will have non-trivial solutions only for certain values of λ^2.)

b) $\dfrac{d^2y}{dx^2} + \lambda^2 y = 1$, $\quad y(0) = 0$, $\quad y(1) = 0$

What if $\lambda = 2n\pi$, $n = 1,2,3,\ldots$?

c) $\dfrac{d^2y}{dx^2} + 4\pi^2 y = 0$, $\quad y(0) = 0$, $\quad y(1) = 1$

5. Find the general solution of each of the following equations.

a) $\dfrac{d^4y}{dx^4} - \lambda^4 y = 0$

b) $y^{(3)} + 3y'' + 3y' + y = 0$, where $y^{(3)} = \dfrac{d^3y}{dt^3}$

c) $y^{(3)} + 2y'' - y' - 2y = t$

(Hint: one characteristic root is $r = 1$)

1.4 NUMERICAL SOLUTIONS

The linear differential equations that have been presented in the previous sections are very fundamental to many applications. A lot is known about them mathematically and many real world problems are adequately represented by them. Non-linear differential equations, on the other hand, present quite a different situation. In many cases they represent the physical problem much better, but are much more difficult to solve exactly. An example of this situation is presented by appropriately representing a population growth problem. If $x(t)$ represents the size of an isolated population, then

$$\frac{dx}{dt} = \alpha\, x(t), \qquad \alpha > 0 \tag{1}$$

$$x(0) = x_0$$

is a linear problem that might represent the population growth under certain circumstances. On the other hand, if food supplies or space allotments are limited, then the exponential growth, as exhibited by the solution to Eq. (1), is not realistic. In this case the non-linear problem

$$\frac{dx}{dt} = \alpha\, x(t) - \beta\, x^2(t), \qquad \alpha > 0, \ \beta > 0 \tag{2}$$

$$x(0) = x_0$$

is a better model to use. Equation (2) is non-linear because of the presence of the $x^2(t)$ term.

The general first order differential equation is

$$\frac{dx}{dt} \;=\; f(t,x) \;, \tag{3}$$

where $x(t)$ is the unknown function of t. Equation (3) is linear only when

$$f(t,x) = a(t)\,x(t) + b(t), \tag{4}$$

and is called non-linear otherwise. The exact solution of non-linear equations is impossible except for some very special types of equations. It is the intent of this section not to discuss these special cases but to give a brief introduction to numerical methods, which is one means of seeking approximate solutions to Eq. (3) when exact solutions are not possible. This use of numerical methods is particularly pertinent with the availability of today's digital computers.

There are many numerical methods used to solve differential equations, but only a few of the most widely used and easily understood will be presented here. Numerical methods apply to initial value problems:

$$\frac{dx}{dt} \;=\; f(t,x)$$

$$\tag{5}$$

$$x(t_0) \;=\; x_0$$

rather than to just the differential equation and will produce a sequence of pairs of points $(t_n,\, x_n)$, n = 1,2,..., that will approximately satisfy Eq. (5). The

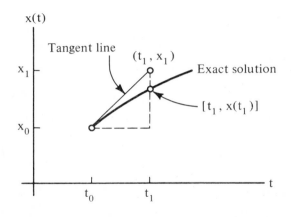

Figure 1.

simplest method, the Euler or tangent line method, can be derived using a geometric approach. From Figure 1. we have that

$$x_1 = x_0 + (t_1 - t_0) f(t_0, x_0) \tag{6}$$

since the slope of the tangent line at (t_0, x_0) is given by

$$\frac{dx}{dt}(t_0) = f(t_0, x_0). \tag{7}$$

Once x_1 is known then $f(t_1, x_1)$ can be computed to give an approximate slope for the solution at t_1. Thus

$$x_2 = x_1 + (t_2 - t_1) f(t_1, x_1) \tag{8}$$

is the approximate solution at $t = t_2$, as shown in Figure 2. This procedure can

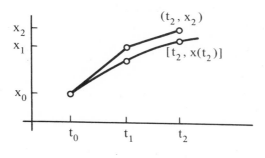

Figure 2.

be continued to yield the general formula

$$x_{n+1} = x_n + (t_{n+1} - t_n) f(t_n, x_n) \tag{9}$$

for the approximate solution at $t = t_{n+1}$. The differences $t_{n+1} - t_n$ are usually taken to be equal so that $t_{n+1} - t_n = h$ for all n, where h is known as the step or mesh size. Thus Eq. (9) becomes

$$x_{n+1} = x_n + h f(t_n, x_n), \tag{10}$$

which is known as the Euler Formula.

The following example will illustrate the use of Eq. (10) while at the same time illustrate some of the results typical to numerical methods.

Example 1. Find the approximate solution to

$$\frac{dx}{dt} = 5x - x^2$$

(11)

$$x(0) = 1,$$

using a step size of 0.1 .

From Eq. (10) we obtain for n = 0

$$x_1 = x_0 + .1 (5x_0 - x_0^2)$$

$$= 1.4$$

since x_0 is given as 1 in the initial conditions and $f(t,x) = 5x - x^2$ from the differential equation. Now that x_1 is "known", Eq. (10) can be used for n = 1:

$$x_2 = x_1 + .1(5x_1 - x_1^2)$$

$$= 1.4 + .1(5(1.4) - (1.4)^2)$$

$$= 1.904,$$

which represents the approximation to x(.2). Continuing in the above fashion, Table I can be constructed to present the pertinent data for determining the approximations x_{n+1} to the exact values $x(t_{n+1})$.

Table I

Numerical Solution Of $x' = 5x - x^2$, x(0) = 1

Using The Euler Formula

n	t_n	x_n	$f(t_n, x_n)$	$hf(t_n, x_n)$	x_{n+1}	exact	% error
0	0	.1	4	.4	1.4	1.459	4%
1	.1	1.4	5.04	.504	1.904	2.023	5.9%
2	.2	1.904	5.89	.589	2.493	2.642	5.6%
3	.3	2.493	6.24	.624	3.117	3.244	3.9%

The exact solution in this case was calculated using techniques not shown in this text and is given here for purposes of comparison. Using a computer and Eq. (10), the above table can be extended very easily.

Table I shows several results typical to numerical techniques. First of all, the "solution" is given in terms of numerical data rather than functions such as in previous sections. Secondly, the errors involved, when compared to the exact solution, are rather complicated. The error as shown involves both local errors and round-off errors. Local errors are caused by using the tangent line rather than the exact curve in going from t_n to t_{n+1}. Round-off errors are due to keeping only a few of the digits in the computations. Using more digits would improve the round-off errors, but even computers only use a finite number of digits. The errors shown are actually the accumulated errors involving both local and round-off errors as well as errors from the preceding steps. In the above example, the accumulated error actually decreases, which is somewhat unusual, due to an inflection point in the curve of the exact solution.

There are numerous ways of decreasing the local errors. One way would be to use a smaller mesh size, thereby not deviating as far from the actual solution curve. This causes problems in the number of calculations and round-off errors. A more pertinent tactic would be to use a more accurate approximation than Eq. (10). Two such approximations will be briefly described here and applied to the previous example for comparison.

The first improvement is given by the improved Euler method. With this method we use Eq. (10) to estimate the value for x_{n+1}, calculate the slope at that point and then average that slope with the slope at x_n to calculate a revised value for x_{n+1}. Writing this in equation form we have

$$x_{n+1} = x_n + h \frac{f(t_n, x_n) + f(t_{n+1}, x^*_{n+1})}{2} \, , \tag{12}$$

where

$$x^*_{n+1} = x_n + hf(t_n, x_n) \tag{13}$$

and $f(t_n, x_n)$ is the same as in Eq. (10) or Eq. (5). A geometrical example of when Eq. (12) is an improvement over Eq. (10) can be seen in Figure 3. for $n = 0$.

A second improved formula is provided by the four-point Runge-Kutta formula:

$$x_{n+1} = x_n + \frac{h}{6} (k_{n,0} + 2k_{n,1} + 2k_{n,2} + k_{n,3}), \tag{14}$$

$$k_{n,0} = f(t_n, x_n)$$

$$k_{n,1} = f(t_n + \tfrac{1}{2}h, x_n + \tfrac{1}{2}h \, k_{n,0})$$

$$k_{n,2} = f(t_n + \tfrac{1}{2}h, x_n + \tfrac{1}{2}h \, k_{n,1}) \tag{15}$$

$$k_{n,3} = f(t_n + h, x_n + h \, k_{n,2}).$$

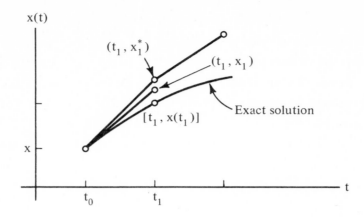

Figure 3.

A careful study of the formulas in Eqs. (14) and (15) shows that the term
$\frac{1}{6}$ ($k_{n,0}$ + $2k_{n,1}$ + $2k_{n,2}$ + $k_{n,3}$) can be interpreted as a weighted average of slopes.
Note that $k_{n,0}$ is the slope at t_n, $k_{n,1}$ is the slope midway between t_n and t_{n+1}
using Eq. (10) to go from t_n to $t_n + \frac{1}{2}h$; $k_{n,2}$ is a second approximation at the
midpoint and $k_{n,3}$ is the slope at the right hand end point t_{n+1}, using $k_{n,2}$ as the
slope in Eq. (10).

   A more detailed analysis of the three numerical techniques given here would
show that the local error caused by the approximation (10) is of the order of h^2
(where h is the mesh size) while that due to approximation (12) is of the order
of h^3 and that of approximation (14) is of the order of h^5. Thus if h is less
than 1, the local error has certainly been improved by the use of either of the
approximations (12) or (14). Table II presents a comparison of the improvements
when the three approximations are used to solve Eq. (11) of Example 1.

Table II

A Comparison of Numerical Solutions to

$x' = 5x - x^2$, $x(0) = 1$, Using a Mesh Size of 0.1.

t_n	Euler	Improved Euler	Runge–Kutta	Exact
0	1	1	1	1
.1	1.400	1.452	1.459	1.459
.2	1.904	2.008	2.021	2.023
.3	2.493	2.620	2.640	2.642
.4	3.117	3.217	3.242	3.244

Further improvements can be obtained either by using a smaller step size or a more accurate method which has not been presented here.

The above discussion has been a brief introduction to the techniques of numerical solution of first order differential equations. The ideas presented here can be extended to solve second and higher order differential equations with both initial value and boundary values specified. These are not presented here, however, as they are best left to a complete course in numerical analysis.

Exercises

1. Rework Example 1. using the Euler method and a step size of .05. Compare your results to the approximate and exact values listed in Table I.

2. Rework Example 1. using the improved Euler method and the Runge-Kutta method with a step size of .05. Do only four lines in a table such as Table I and compare your results with those shown in Table II.

3. Use the improved Euler formula to find an approximation to $y(.3)$ when

    $\frac{dy}{dx} = 1 + x + 2y$, $y(0) = 1$. Use a step size of 0.1 and compare your result

    to the exact value. (The initial value problem can be solved exactly using methods of Section 1.1.).

4. Use all three methods to find an approximation to $x(.2)$ when $\frac{dx}{dt} = t + x^2$,

    $x(0) = 1$ using a step size of 0.1.

5. Use the Runge-Kutta method to find an approximation to $x(2)$ when $\frac{dx}{dt} = x^2 e^{-t}$,

    $x(1) = 2$ using a step size of 0.5. Repeat using a step size of 0.2 and compare results. Can you conclude anything about the accuracy of your results?

6. Use the Euler method to find an approximation to $y(1)$ when $\frac{dy}{dx} = \frac{x}{y}$, $y(0) = 1$

    using a step size of 0.1. Compare your results with the exact value as given by $y^2 = 1 + x^2$. How does the error vary as x increases?

2

MATRICES AND
LINEAR EQUATIONS

2.1 MATRIX ALGEBRA

Many management, economic and engineering problems require the solution of system of <u>linear algebraic equations</u>. For instance a marketing situation gives rise to the system of equations shown in the following example:

Example 1.

$$.6x_1 + .2x_2 + .2x_3 = a_1$$

$$.5x_1 + .3x_2 + .2x_3 = a_2 \tag{1}$$

$$.4x_1 + .2x_2 + .4x_3 = a_3$$

where a_1, and a_2 and a_3 are known quantities and it is desired to determine x_1, x_2 and x_3.

This is an example of three equations in three unknowns. Many practical problems, however, would lead to more equations and more unknowns. Matrices are used in conjunction with this type of problem for several reasons: the notation is simplified, theoretical results are obtained more easily and practical results, especially on computers, are heavily dependent on the matrix formulation. It is the purpose of this chapter to introduce the concepts involved in matrix manipulations pertaining to various ways of solving and interpreting systems of equations.

For the purposes of this text a matrix may be considered to be a rectangular array of real numbers. The array is enclosed by square brackets, as illustrated in the following example.

Example 2.
$$A = \begin{bmatrix} 3 & 2 & 1 & -1 \\ 10 & 1 & -3 & 2 \\ 1 & 0 & 5 & -5 \end{bmatrix} \tag{2}$$

The matrix of Example 2. is said to be a 3 × 4 matrix since it has 3 rows and 4 columns. If the number of rows equals the number of columns, then the matrix is called a square matrix.

The elements of any matrix A are symbolized by a_{ij} with i denoting the row and j the column in which a_{ij} appears. Hence in the matrix of Example 2, $a_{21} = 10$, since 10 is the element in the second row and first column, and $a_{34} = -5$, since -5 is the element in the third row fourth column. We often use A and $[a_{ij}]$ as interchangeable symbols for the same matrix. We shall use both notations in the following definitions of operations on matrices and relations between matrices.

Two matrices A and B are equal when $a_{ij} = b_{ij}$ for all i and j. That is, all the elements are equal.

Example 3.

$$\text{If} \quad A = \begin{bmatrix} x & 5 \\ -1 & 2 \end{bmatrix}, \quad B = \begin{bmatrix} 1 & 5 \\ -1 & 2 \end{bmatrix} \quad \text{and} \quad C = \begin{bmatrix} 1 & 5 & 1 \\ -1 & 2 & 0 \end{bmatrix},$$

then the matrices A and B are equal only when x = 1, while C cannot be equal to either A or B since it is not the right size (C is 2 × 3 while A and B are 2 × 2).

Two matrices may be added if and only if they are the same size. In this case the sum of two matrices A and B is a third matrix C, also the same size as A and B, whose elements are the sum of the corresponding elements of A and B. If we designate C by

$$C = A + B,$$

then the elements are related by

$$c_{ij} = a_{ij} + b_{ij}. \tag{3}$$

Example 4.

$$\text{If} \quad A = \begin{bmatrix} 3 & 5 & -1 \\ 0 & 2 & 1 \end{bmatrix} \quad \text{and} \quad B = \begin{bmatrix} 1 & 0 & 1 \\ 1 & -1 & 2 \end{bmatrix}$$

then

$$A + B = C = \begin{bmatrix} 4 & 5 & 0 \\ 1 & 1 & 3 \end{bmatrix}.$$

The <u>product</u> of a matrix A <u>by a scalar</u> d is defined by the equation

$$dA = [d\ a_{ij}], \tag{4}$$

where the scalar d is simply any real number.

Example 5.

$$\text{If } A = \begin{bmatrix} 1 & 5 \\ 0 & 2 \\ -1 & 1 \end{bmatrix}, \text{ then } 3A \text{ is given by } \begin{bmatrix} 3 & 15 \\ 0 & 6 \\ -3 & 3 \end{bmatrix}.$$

The product defined by Eq. (4) is often referred to as <u>scalar</u> multiplication. It can be used, together with addition, to accomplish subtraction since one may first multiply B by -1 and then add to A to find $A - B$.

The <u>product</u> <u>of</u> <u>two</u> <u>matrices</u> is defined by the following procedure: Let $A = [a_{ij}]$ be an $n \times s$ matrix and $B = [b_{ij}]$ be an $s \times m$ matrix. Then $AB = C$, where $C = [c_{ij}]$ is the $n \times m$ matrix with elements defined as

$$c_{ij} = \sum_{k=1}^{s} a_{ik}\ b_{kj}. \tag{5}$$

Before looking at an example, let's "read" what the definition says. The element appearing in the i, j position in C is the sum of the products of the elements appearing in the ith row of A times the corresponding elements appearing in the jth column of B. Notice that for the multiplication of matrices to be defined the number of columns of A, s, and the number of rows of B, s, must be the same. The diagram of Figure 1. illustrates this size restriction for the product $AB = C$ to be defined.

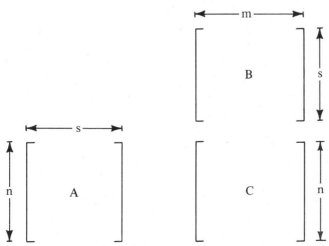

Figure 1.

Example 6.

Find AB if A $= \begin{bmatrix} 1 & 0 \\ 4 & 3 \\ 2 & 1 \end{bmatrix}$ and B $= \begin{bmatrix} 1 & 0 \\ 0 & 2 \end{bmatrix}$.

$$\begin{bmatrix} 1 & 0 \\ 4 & 3 \\ 2 & 1 \end{bmatrix} \begin{bmatrix} 1 & 0 \\ 0 & 2 \end{bmatrix} = \begin{bmatrix} 1 \cdot 1 + 0 \cdot 0 & 1 \cdot 0 + 0 \cdot 2 \\ 4 \cdot 1 + 3 \cdot 0 & 4 \cdot 0 + 3 \cdot 2 \\ 2 \cdot 1 + 1 \cdot 0 & 2 \cdot 0 + 1 \cdot 2 \end{bmatrix}$$

$$= \begin{bmatrix} 1 & 0 \\ 4 & 6 \\ 2 & 2 \end{bmatrix}$$

Notice in the example how the rows of A multiply the columns of B.

The motivation leading to the above definition of the product of two matrices is beyond the scope of this text. The relationship given in Eq. (5) is a generalization of the concepts developed in Exercise 8. at the end of this section, where the definition (5) is developed for n, m and s all equal to 2.

The definition of multiplication can be used to condense sets of equations such as we saw in Example 1. This procedure is demonstrated in the following example.

Example 7.   Write the equations below in matrix notation:

$$2x_1 + 3x_2 = 1$$

$$x_1 + 2x_2 = 5 .$$

(6)

If we let A $= \begin{bmatrix} 2 & 3 \\ 1 & 2 \end{bmatrix}$, X $= \begin{bmatrix} x_1 \\ x_2 \end{bmatrix}$ and B $= \begin{bmatrix} 1 \\ 5 \end{bmatrix}$,

then

$$AX = \begin{bmatrix} 2x_1 + 3x_2 \\ x_1 + 2x_2 \end{bmatrix} .$$

Thus, by using the definition of equality, the given equations can be written as

$$AX = B. \tag{7}$$

Clearly by letting A be a general $n \times n$ and X and B be general $n \times 1$ matrices then Eq. (7) also represents n equations in n unknowns.

The following <u>algebraic laws</u> of matrix multiplication hold:

1) Associative: $A(B\,C) = (A\,B)C,$ $\tag{8}$

2) Distributive: $A(B + C) = AB + AC,$ $\tag{9}$

but the commutative law does <u>not</u> hold:

3) $AB \neq BA$ $\tag{10}$

Equations (8) and (9) can be verified using definitions given above. The failure of commutativity is easily verified since one of the products need not be defined. Even if both products are defined, the products may be of different sizes; and even if both products are defined and of the same size, they may still be different, as the following example shows:

Example 8.

$$\text{If} \quad A = \begin{bmatrix} 1 & 2 \\ 3 & 4 \end{bmatrix} \quad \text{and} \quad B = \begin{bmatrix} 2 & 1 \\ -1 & 0 \end{bmatrix},$$

then

$$AB = \begin{bmatrix} 0 & 1 \\ 2 & 3 \end{bmatrix} \quad \text{and} \quad BA = \begin{bmatrix} 5 & 8 \\ -1 & -2 \end{bmatrix}$$

so that $AB \neq BA$.

There are, however, important instances for which $AB = BA$.

The <u>transpose</u> of a matrix A is the matrix A^T obtained from A by interchanging the rows and columns of A. That is the element c_{ij} of A^T is a_{ji}.

Example 9.

$$\text{If} \quad A = \begin{bmatrix} 1 & 2 & 3 \\ 4 & 5 & 6 \end{bmatrix} \quad \text{then} \quad A^T = \begin{bmatrix} 1 & 4 \\ 2 & 5 \\ 3 & 6 \end{bmatrix}.$$

If A is $n \times m$ then A^T is $m \times n$. The transpose of a product is the product of the transposes in <u>reverse</u> order. That is

$$(AB)^T = B^T A^T. \tag{11}$$

The elements a_{ii} of the matrix A are called the <u>diagonal elements</u>. If A is a square matrix where $a_{ij} \doteq 0$ when $i \neq j$ then it is called a <u>diagonal matrix</u>. A very important diagonal matrix is the identity matrix I. I is a square $(n \times n)$ matrix where $I = (\delta_{ij})$ and

$$\delta_{ij} = \begin{cases} 0 & i \neq j \\ 1 & i = j \end{cases}. \tag{12}$$

The element δ_{ij} is known as the <u>Kronecker delta</u>, which appears in a variety of applications other than with matrices. Using Eq. (12), we find that for a 4×4 matrix I has the following form:

$$I = \begin{bmatrix} 1 & 0 & 0 & 0 \\ 0 & 1 & 0 & 0 \\ 0 & 0 & 1 & 0 \\ 0 & 0 & 0 & 1 \end{bmatrix}. \tag{13}$$

Thus the reader can easily verify that

$$IB = BI = B \tag{14}$$

for any matrix B for which the products are defined.

A <u>triangular matrix</u> is a square matrix whose elements above the diagonal or below the diagonal are all zero. Triangular matrices play an important role in matrix computations as they are much less cumbersome to use.

Example 10.

$$A = \begin{bmatrix} 1 & 6 & -1 \\ 0 & 5 & 2 \\ 0 & 0 & 4 \end{bmatrix} \quad \text{and} \quad B = \begin{bmatrix} 1 & 0 & 0 \\ -3 & 1 & 0 \\ 2 & 0 & 5 \end{bmatrix}$$

are both triangular matrices.

In fact, a diagonal matrix is also a special case of a triangular matrix.

A very crucial question can now be raised: Is there a matrix C such that

$$CA = I \tag{15}$$

for a given square matrix A? The answer depends on A, but if a matrix C does
exist that satisfies Eq. (15), then C is called the inverse of A and is denoted
as $C = A^{-1}$. It should be noted that the inverse matrix, A^{-1}, is used rather than
division by the matrix A, since <u>division by a matrix is not defined</u>. If A^{-1}
exists, then its role as a "divisor" is demonstrated in the following example of the
symbolic solution of a system of equations.

Example 11.   Find the solution of

$$AX = B. \tag{16}$$

If the inverse exists, then multiply both sides of Eq. (16) by A^{-1} to
get

$$A^{-1}(AX) = A^{-1}B,$$

which becomes

$$(A^{-1}A)X = A^{-1}B \tag{17}$$

by the associative law, Eq. (8). But $A^{-1}A = I$ by Eq. (15), so that
we obtain

$$IX = X = A^{-1}B \tag{18}$$

as the solution to Eq. (16).

We say that Eq. (16) is symbolically solved by Eq. (18) since it is no easy matter
(and rather inefficient also) to find A^{-1}, as is demonstrated by the development of
the next two sections.

Exercises

1. Let $A = \begin{bmatrix} 1 & 3 & -2 \\ 0 & -1 & -3 \end{bmatrix}$ and $B = \begin{bmatrix} -2 & 1 & 0 \\ 3 & -3 & 5 \end{bmatrix}$.

   a) Find A + B   b) Find 2A - 3B

   c) Find A^T d) Find a matrix C such that
                                              A + C = B

2. Find matrices A, B and X so that AX = B represents the system of equations

$$x_1 + 3x_2 - x_3 + x_4 = 1$$

$$2x_1 - x_2 + x_4 = 3$$

$$5x_1 + x_2 + 9x_3 = 5 .$$

3. Let $A = [a_{ij}]$ be a 3 × t matrix and $B = [b_{ij}]$ be a 4 × 5 matrix.

a) Under what conditions does AB exist?

b) What size is AB if it exists?

c) Does BA ever exist? Why?

4. Suppose that A is 3 × 1, B is 3 × 2 and C is 3 × 3, which of the following operations are possible?

a) AA^T b) AC   c) CA   d) AB

e) A^TCB f) $A^TB + C$ g) $A + A^T$ h) CC^T

5. If $A = \begin{bmatrix} 2 & 3 & 4 & 4 \\ 1 & 0 & -1 & 6 \\ 0 & 1 & 2 & 9 \end{bmatrix}$ and $B = \begin{bmatrix} 0 & 2 \\ 3 & 1 \\ 1 & 0 \\ 0 & -1 \end{bmatrix}$, find AB.

6. Let $A = \begin{bmatrix} 9 & 6 & 2 \\ 4 & 3 & 1 \end{bmatrix}$ and $B = \begin{bmatrix} 2 & 4 \\ 0 & 9 \\ 0 & 0 \end{bmatrix}$.

a) Find AB and BA

b) Find $(AB)^T$ and B^TA^T

7. Show that B is the inverse of the matrix A, where

$$A = \begin{bmatrix} 1 & 0 & -1 \\ 0 & 1 & 2 \\ 2 & 3 & 5 \end{bmatrix} \quad \text{and} \quad B = \begin{bmatrix} -1 & -3 & 1 \\ 4 & 7 & -2 \\ -2 & -3 & 1 \end{bmatrix}.$$

Use this fact to solve the system of equations

$$x_1 \qquad\qquad - \quad x_3 \;=\; 2$$

$$x_2 \;+\; 2x_3 \;=\; 1$$

$$2x_1 \;+\; 3x_2 \;+\; 5x_3 \;=\; -1\;.$$

8. To motivate the matrix multiplication definition consider three coordinate systems in the plane, denoted as the x_1x_2 system, the y_1y_2 system and the z_1z_2 system. If the systems are related to each other by rotations about the origin, then the equations

$$x_1 \;=\; a_{11}y_1 \;+\; a_{12}y_2$$

$$x_2 \;=\; a_{21}y_1 \;+\; a_{22}y_2 \quad, \tag{1}$$

$$y_1 \;=\; b_{11}z_1 \;+\; b_{12}z_2$$

$$y_2 \;=\; b_{21}z_1 \;+\; b_{22}z_2 \quad, \tag{2}$$

and

$$x_1 \;=\; c_{11}z_1 \;+\; c_{12}z_2$$

$$x_2 \;=\; c_{21}z_1 \;+\; c_{22}z_2 \tag{3}$$

relate the coordinates in the various systems. Find formulas for the c_{ij}'s in terms of the a_{ij}'s and the b_{ij}'s by substituting Eqs. (2) into Eqs. (1) and comparing the results with Eqs. (3). You should obtain Eq. (5) with s = 2. Finally write Eqs. (1), (2) and (3) in matrix notation and verify that Eq. (5) of the text does indeed represent the product.

2.2 DETERMINANTS AND THE MATRIX INVERSE

Up to this point addition, subtraction and multiplication of matrices have been discussed. "Division" of matrices is done through the matrix inverse, as was discussed briefly at the end of the previous section. Two methods for finding the inverse of a matrix are: the adjoint method and the elementary row operation method. The adjoint method will be developed here and the second method in the next section.

The adjoint method for finding an inverse of a matrix requires the evaluation of determinants, which will be reviewed first. For any <u>square</u> <u>matrix</u> A there is a quantity called the determinant of A, or briefly det A. The notation used for determinants is shown in Eq. (1):

$$\det A = \begin{vmatrix} a_{11} & a_{12} & a_{13} \\ a_{21} & a_{22} & a_{23} \\ a_{31} & a_{32} & a_{33} \end{vmatrix} \tag{1}$$

where A is a 3 × 3 matrix. The definition of a 2 × 2 determinant is

$$\begin{vmatrix} a_{11} & a_{12} \\ a_{21} & a_{22} \end{vmatrix} = a_{11} a_{22} - a_{12} a_{21} \ . \tag{2}$$

In Eq. (2) we see the important fact that a determinant is a real (scalar) number that is <u>associated</u> with the given array.

Example 1.

$$\text{Evaluate} \begin{vmatrix} 1 & 3 \\ 2 & 5 \end{vmatrix} \ .$$

$$\begin{vmatrix} 1 & 3 \\ 2 & 5 \end{vmatrix} = 1 \cdot 5 - 3 \cdot 2 = -1,$$

by reference to Eq. (2).

The extension of Eq. (2) to 3 × 3 or n × n determinants is quite complicated, for there are many more products and "permutations" of subscripts to consider. We shall derive a more practical result for n × n determinants which also introduces concepts that are applicable in a variety of problems.

We are discussing the general $n \times n$ determinant:

$$\det A = \begin{vmatrix} a_{11} & a_{12} & \cdots & a_{1n} \\ a_{21} & & & \\ \cdot & & & \\ \cdot & & & \\ \cdot & & & \\ a_{n1} & \cdots & \cdots & a_{nn} \end{vmatrix} . \qquad (3)$$

For the a_{ij} element, the $(n-1) \times (n-1)$ determinant formed by omitting the ith row and the jth column of det A is called the <u>minor</u> of a_{ij}. The minor of a_{ij} multiplied by $(-1)^{i+j}$ is called the <u>cofactor</u> of a_{ij}.

Example 2.

$$\text{For } \det A = \begin{vmatrix} 1 & 3 & 5 \\ 0 & -1 & 4 \\ 9 & 6 & 5 \end{vmatrix} , \text{ find the minor and}$$

cofactor for each of the elements a_{22} and a_{23}.

The minor of a_{22} is the determinant $\begin{vmatrix} 1 & 5 \\ 9 & 5 \end{vmatrix}$ and its cofactor is the

same since $(-1)^{2+2} = (-1)^4 = +1$. The minor of a_{23} is $\begin{vmatrix} 1 & 3 \\ 9 & 6 \end{vmatrix}$ and

its cofactor is $-\begin{vmatrix} 1 & 3 \\ 9 & 6 \end{vmatrix}$ since $(-1)^{2+3} = (-1)^5 = -1$.

Clearly an $n \times n$ determinant has n^2 minors or cofactors.

The value of an n × n determinant can now be defined by the following expression:

$$\text{Det } A = \sum_{j=1}^{n} a_{ij} \text{ cof } (a_{ij}) = \sum_{i=1}^{n} a_{ij} \text{ cof } (a_{ij}). \tag{4}$$

If i is fixed (as in the middle member of Eq. (4)) then the expression is called the expansion by the ith row. If j is fixed (as in the right hand member) then the expression is called the expansion by the jth column. Equation (4) will suffice for our purposes, although it is not the actual mathematical definition of a determinant. In evaluating Eq. (4) it is important to realize that <u>any</u> row or column used will give the same results, since for the middle term there is no restriction on which i is used and in the right hand term there is no restriction on which j is used. The following example will demonstrate the use of Eq. (4).

Example 3.

$$\text{Evaluate } \begin{vmatrix} 0 & 1 & 2 \\ -1 & 3 & 0 \\ 1 & -2 & 1 \end{vmatrix}, \text{ using the first and then the}$$

second row.

Using the first row we have:

$$\begin{vmatrix} 0 & 1 & 2 \\ -1 & 3 & 0 \\ 1 & -2 & 1 \end{vmatrix} = (0)(-1)^{1+1} \begin{vmatrix} 3 & 0 \\ -2 & 1 \end{vmatrix} + 1\,(-1)^{1+2} \begin{vmatrix} -1 & 0 \\ 1 & 1 \end{vmatrix}$$

$$+ 2\,(-1)^{1+3} \begin{vmatrix} -1 & 3 \\ 1 & -2 \end{vmatrix}$$

$$= 0\,(3 - 0) - (-1 - 0) + 2\,(2 - 3) = -1.$$

For the expansion using the second row we have:

$$\begin{vmatrix} 0 & 1 & 2 \\ -1 & 3 & 0 \\ 1 & -2 & 1 \end{vmatrix} = -1(-1)^{2+1} \begin{vmatrix} 1 & 2 \\ -2 & 1 \end{vmatrix} + 3\,(-1)^{2+2} \begin{vmatrix} 0 & 2 \\ 1 & 1 \end{vmatrix}$$

$$+ 0\,(-1)^{2+3} \begin{vmatrix} 0 & 1 \\ 1 & -2 \end{vmatrix}$$

$$= -1\,(-1)(1 + 4) + 3\,(-2) + 0\,(-1)(-1) = -1.$$

The minors and cofactors have been explicitly shown to help with the understanding of the definition. With practice the sign convention has a noticeable pattern and is not hard to remember. Notice how use is made of the zeros occuring in the determinant by picking rows in which they appear. The student should evaluate the same determinant using other rows or columns.

To evaluate a 4×4 determinant, the cofactors involved as shown in Eq. (4) are 3×3 determinants, which in turn have to be evaluated as shown in Example 3. Clearly this becomes rather tedious and lengthy (even for high speed computers). For this reason we introduce the following properties of determinants:

Property 1 If A is a diagonal or a triangular matrix with diagonal elements a_{ii}, $i = 1,2,\ldots n$, then $\det A = a_{11}a_{22} \cdots a_{nn}$.

Property 2 If B is obtained from A by interchanging any two rows of A, then $\det B = - \det A$.

Property 3 If B is obtained from A by multiplying all the elements of any row of A by c, then $\det B = c \det A$.

Property 4 If all the elements of any two rows of A are proportional, then $\det A = 0$.

Property 5 If B is obtained from A by replacing the ith row of A by the row of elements composed of the original i^{th} row elements added to k times the corresponding elements of any other row of A, then $\det A = \det B$.

All of these properties may be verified using Eq. (4); and hence any place where "row" appears may be changed to "column" without affecting the result.

Properties 2, 3 and 5 introduce what are called the elementary row operations. By the appropriate use of these row operations we will be able not only to simplify the evaluation of a determinant but also to simplify the finding of the inverse of a matrix and the solving of a system of equations, as shown in the next two sections. Example 4. demonstrates how the above properties are utilized in the evaluation of the same determinant that appeared in Example 3.

Example 4.

$$\text{Evaluate} \quad \begin{vmatrix} 0 & 1 & 2 \\ -1 & 3 & 0 \\ 1 & -2 & 1 \end{vmatrix}, \text{ using the above properties.}$$

$$\begin{vmatrix} 0 & 1 & 2 \\ -1 & 3 & 0 \\ 1 & -2 & 1 \end{vmatrix} = - \begin{vmatrix} 1 & -2 & 1 \\ -1 & 3 & 0 \\ 0 & 1 & 2 \end{vmatrix}$$ by Prop. 2, since the first and third rows have been interchanged.

$$= - \begin{vmatrix} 1 & -2 & 1 \\ 0 & 1 & 1 \\ 0 & 1 & 2 \end{vmatrix}$$ by Prop. 5, since one times the first row has been added to the second row.

$$= - \begin{vmatrix} 1 & -2 & 1 \\ 0 & 1 & 1 \\ 0 & 0 & 1 \end{vmatrix}$$ by Prop. 5, since (-1) times the second row has been added to the third row.

$$= - 1 .$$ by Prop. 1, since the last determinant is triangular.

As illustrated in Example 4, the <u>object</u> is to use Property 5 <u>to find a triangular determinant</u> <u>which</u> <u>is a multiple</u> <u>of the original</u> determinant. Properties 2 and 3 are used here basically to simplify the hand calculations. The saving in Example 4. was perhaps not obvious, but for 4 × 4 and larger determinants the use of Eq. (4) is very impractical as compared to using Property 5.

We are now in a position to calculate the inverse of a matrix. Let $\operatorname{cof}(a_{ij})$ denote the cofactor of the element of a_{ij} in the matrix A. Then the matrix $C^T = [c_{ji}]$ is called the <u>adjoint</u> of A and is denoted as adj A. That is

$$\text{adj } A = C^T = (c_{ji}) \tag{5}$$

where

$$c_{ij} = \operatorname{cof}(a_{ij}). \tag{6}$$

It can be shown that

$$(\text{adj } A)A = A(\text{adj } A) = (\det A)I, \tag{7}$$

and hence if $\det A \neq 0$

$$\frac{\text{adj } A}{\det A} A = I. \tag{8}$$

Equation (8) implies that

$$A^{-1} = \frac{\text{adj } A}{\det A} , \tag{9}$$

based upon Eq. (14), Section 2.1, and the discussion following it. If det A = 0, then Eq. (8) no longer holds and thus there is no matrix B for which BA = I, and hence no inverse exists.

Example 5.   Find the adjoint and inverse of

$$A \;=\; \begin{bmatrix} 0 & 1 & 2 \\ -1 & 3 & 0 \\ 1 & -2 & 1 \end{bmatrix} .$$

There are nine cofactors to be found. The first three are:

$$c_{11} \;=\; \begin{vmatrix} 3 & 0 \\ -2 & 1 \end{vmatrix} \;=\; 3, \qquad c_{12} \;=\; - \begin{vmatrix} -1 & 0 \\ 1 & 1 \end{vmatrix} \;=\; 1$$

and

$$c_{13} \;=\; \begin{vmatrix} -1 & 3 \\ 1 & -2 \end{vmatrix} \;=\; -1.$$

The rest are found in a similar way to yield

$$C \;=\; \begin{bmatrix} 3 & 1 & -1 \\ -5 & -2 & 1 \\ -6 & -2 & 1 \end{bmatrix} .$$

Hence the adjoint of A is given by

$$\text{adj } A = C^T \;=\; \begin{bmatrix} 3 & -5 & -6 \\ 1 & -2 & -2 \\ -1 & 1 & 1 \end{bmatrix}$$

and the inverse of A is given by

$$A^{-1} \;=\; \frac{\text{adj } A}{\det A} \;=\; \begin{bmatrix} -3 & 5 & 6 \\ -1 & 2 & 2 \\ 1 & -1 & -1 \end{bmatrix}$$

since det A = -1 as found in Example 4.

When calculating the inverse of a matrix it is always possible to check the result by calculating the product $A^{-1} A$, which should be done. The student may verify the answer to the above example in this fashion.

Exercises

1. Use Eq. (4) to evaluate the determinant

$$\begin{vmatrix} 3 & -1 & 1 & -2 \\ 1 & 2 & 2 & 3 \\ 1 & 0 & -2 & 0 \\ 4 & -3 & 0 & 2 \end{vmatrix}$$

2. Use Properties 2, 3 and 5 to evaluate the determinant of Exercise 1.

3. Verify that the determinant of a triangular matrix is $a_{11} a_{22} \cdots a_{nn}$.

4. Using the result of Exercise 3, evaluate

$$\begin{vmatrix} 1 & -10 & 100 & 3 \\ 0 & 3 & 99 & 6 \\ 0 & 0 & 2 & 1 \\ 0 & 0 & 0 & -1 \end{vmatrix}.$$

5. Does $\begin{bmatrix} 1 & 3 & 5 \\ 0 & 1 & 2 \\ 0 & 0 & 0 \end{bmatrix}$ have an inverse? Explain.

6. Let A $= \begin{bmatrix} 1 & 1 & 0 \\ 1 & 1 & 1 \\ 0 & 2 & 1 \end{bmatrix}$.   Find adj A and A^{-1}.

7. Let A $= \begin{bmatrix} 1 & 0 & 0 & 0 \\ 0 & 2 & 0 & 0 \\ 0 & 0 & 3 & 0 \\ 0 & 0 & 0 & -1 \end{bmatrix}$.   Find adj A and A^{-1}.

8. Evaluate each of the following determinants.

a)
$$\begin{vmatrix} 12 & 28 & 40 \\ 5 & 5 & 2 \\ -3 & 9 & 6 \end{vmatrix}$$

b)
$$\begin{vmatrix} 1 & 3 & -1 & 2 \\ 2 & 0 & -1 & -2 \\ -1 & 2 & 3 & 4 \\ 0 & 5 & 2 & 6 \end{vmatrix}$$

c)
$$\begin{vmatrix} -1 & 2 & 3 & 5 \\ 2 & 1 & 0 & -4 \\ 6 & -12 & 2 & -10 \\ 1 & 3 & -1 & -1 \end{vmatrix}$$

9. It can be shown that det (AB) = det A det B. Verify this formula for

$$A \;=\; \begin{bmatrix} 0 & 1 & 2 \\ -1 & 3 & 0 \\ 1 & -2 & 1 \end{bmatrix} \quad \text{and} \quad B \;=\; \begin{bmatrix} 1 & 2 & 2 \\ 2 & -3 & -1 \\ 1 & 0 & 1 \end{bmatrix}.$$

10. Find the formula for the inverse of $A \;=\; \begin{bmatrix} a_{11} & a_{12} \\ a_{21} & a_{11} \end{bmatrix}$ using the adjoint method.

2.3 ROW OPERATIONS AND THE MATRIX INVERSE

In the last section we saw how to compute the matrix inverse using the adjoint matrix approach. For "large" matrices this is rather tedious for there are n^2 determinants of size $(n-1) \times (n-1)$ to evaluate in forming the cofactor matrix. In this section we will introduce the row operation for matrices and find the matrix inverse using row operations.

The following operations are called <u>elementary</u> <u>row</u> <u>operations</u>:
1) interchanging any two rows of a matrix
2) multiplying any row of a matrix by a non-zero scalar
3) adding a scalar multiple of any row of a matrix to any other row of the same matrix.

If a matrix A can be transformed into a matrix B by means of one or more elementary row operations, then we write $A \sim B$ and say that A is _equivalent_ to B.

Example 1.

$$\begin{bmatrix} 1 & 4 & 3 \\ -2 & 3 & 1 \end{bmatrix} \quad \sim \quad \begin{bmatrix} 1 & 4 & 3 \\ 0 & 11 & 7 \end{bmatrix}$$

since the right hand matrix is obtained from the first by multiplying the first row by 2 and adding to the second row.

An _elementary_ _matrix_ is a matrix that can be obtained from the identity matrix I by an elementary row operation.

Example 2.

$$E_1 = \begin{bmatrix} 0 & 1 & 0 \\ 1 & 0 & 0 \\ 0 & 0 & 1 \end{bmatrix} , \quad E_2 = \begin{bmatrix} 1 & 0 & 0 \\ 0 & c & 0 \\ 0 & 0 & 1 \end{bmatrix} , \text{ and } E_3 = \begin{bmatrix} 1 & 0 & 0 \\ 2 & 1 & 0 \\ 0 & 0 & 1 \end{bmatrix}$$

are elementary matrices since each is obtained from I by one of the elementary row operations.

If an elementary matrix multiplies a matrix A on the left then it performs an elementary row operation on A.   This will be illustrated in Example 3.

Example 3.

$$\text{Given } A = \begin{bmatrix} -2 & 1 & 1 \\ 1 & 3 & 1 \\ 4 & -1 & 1 \end{bmatrix} , \text{ then } E_1 A \text{ is given by:}$$

$$\begin{bmatrix} 0 & 1 & 0 \\ 1 & 0 & 0 \\ 0 & 0 & 1 \end{bmatrix} \begin{bmatrix} -2 & 1 & 1 \\ 1 & 3 & 1 \\ 4 & -1 & 1 \end{bmatrix} = \begin{bmatrix} 1 & 3 & 1 \\ -2 & 1 & 1 \\ 4 & -1 & 1 \end{bmatrix} = B,$$

where the matrix B is the same as A with the first and second rows interchanged. Multiplying this new matrix by E_3, we get:

$$\begin{bmatrix} 1 & 0 & 0 \\ 2 & 1 & 0 \\ 0 & 0 & 1 \end{bmatrix} \begin{bmatrix} 1 & 3 & 1 \\ -2 & 1 & 1 \\ 4 & -1 & 1 \end{bmatrix} = \begin{bmatrix} 1 & 3 & 1 \\ 0 & 7 & 3 \\ 4 & -1 & 1 \end{bmatrix} = C,$$

where the matrix   C   is obtained from   B   by multiplying the first row
of   B   by   2   and adding to the second row of   B.

The process illustrated in Example 3. can be written in condensed notation as

$$C = E_3 B = E_3(E_1 A) = (E_3 E_1)A. \tag{1}$$

It should be clear that general n × n elementary matrices will perform the same row
operations on general n × n matrices. The analogies and differences of the elementary row operations and elementary matrices to the Properties 2, 3 and 5 of determinants (Section 2.2) should be considered.

Now, **su**ppose there is a sequence of elementary matrices for a square matrix
A   which will yield the result

$$E_n \cdots E_2 E_1 A = I, \tag{2}$$

where E_1, E_2, E_3 here need not be the same as in Example 2. By multiplying both
sides of Eq. (2) on the right by A^{-1}, we get

$$(E_n \cdots E_2 E_1 A)A^{-1} = IA^{-1} = A^{-1}, \tag{3}$$

which by associativity yields

$$(E_n \cdots E_2 E_1)I = A^{-1}. \tag{4}$$

Comparing Eqs. (2) and (4) we see the result that: <u>If a sequence of elementary row
operations reduces A to the identity matrix, then the same sequence of elementary
row operations performed on the identity matrix will yield the inverse of A</u>. In
practice we do not use the elementary matrices, but simply perform the elementary row
operations simultaneously on A and I to obtain equivalent matrices. This will be
illustrated in Example 4.

Example 4.

$$\text{Find the inverse of } A = \begin{bmatrix} 1 & 0 & 2 \\ 2 & -1 & 3 \\ 4 & 1 & 8 \end{bmatrix}.$$

To do this we form a new matrix that has A and I combined and proceed
as follows:

$$\left[\begin{array}{ccc|ccc} 1 & 0 & 2 & 1 & 0 & 0 \\ 2 & -1 & 3 & 0 & 1 & 0 \\ 4 & 1 & 8 & 0 & 0 & 1 \end{array} \right] \sim \left[\begin{array}{ccc|ccc} 1 & 0 & 2 & 1 & 0 & 0 \\ 0 & -1 & -1 & -2 & 1 & 0 \\ 4 & 1 & 8 & 0 & 0 & 1 \end{array} \right]$$

$$\sim \begin{bmatrix} 1 & 0 & 2 & 1 & 0 & 0 \\ 0 & -1 & -1 & -2 & 1 & 0 \\ 0 & 1 & 0 & -4 & 0 & 1 \end{bmatrix}$$

$$\sim \begin{bmatrix} 1 & 0 & 2 & 1 & 0 & 0 \\ 0 & -1 & -1 & -2 & 1 & 0 \\ 0 & 0 & -1 & -6 & 1 & 1 \end{bmatrix}$$

$$\sim \begin{bmatrix} 1 & 0 & 0 & -11 & 2 & 2 \\ 0 & -1 & 0 & 4 & 0 & -1 \\ 0 & 0 & -1 & -6 & 1 & 1 \end{bmatrix}$$

$$\sim \begin{bmatrix} 1 & 0 & 0 & -11 & 2 & 2 \\ 0 & 1 & 0 & -4 & 0 & 1 \\ 0 & 0 & 1 & 6 & -1 & -1 \end{bmatrix}.$$

Now, since A is equivalent to the identity matrix, seen by comparing
the left portions of the first and last matrix, we thus conclude that

$$A^{-1} = \begin{bmatrix} -11 & 2 & 2 \\ -4 & 0 & 1 \\ 6 & -1 & -1 \end{bmatrix},$$

by comparing the right portions.

The student should verify that $A^{-1} A = I$ for the result of Example 4.

Before proceeding it should be emphasized that by using row operations and
equivalent matrices we have been able to find the inverse of a matrix, as shown
in Example 4. The elementary matrices were used simply as a way of justifying the
results and showing how an equality could be introduced if needed. For computational
work, the elementary matrices are not needed.

In the last lecture, we saw that A^{-1} could not be found if det A = 0. The
astute student might ask whether this same criterion were true using row operations.
The following example will illustrate, using row operations, what happens when A^{-1}
does not exist.

Example 5.

     If it exists, find the inverse of $A = \begin{bmatrix} 1 & 0 & 2 \\ 2 & -1 & 3 \\ 4 & 0 & 8 \end{bmatrix}$.

     Proceeding as before, we obtain

$$\left[\begin{array}{ccc|ccc} 1 & 0 & 2 & 1 & 0 & 0 \\ 2 & -1 & 3 & 0 & 1 & 0 \\ 4 & 0 & 8 & 0 & 0 & 1 \end{array}\right] \sim \left[\begin{array}{ccc|ccc} 1 & 0 & 2 & 1 & 0 & 0 \\ 0 & -1 & -1 & -2 & 1 & 0 \\ 0 & 0 & 0 & -4 & 0 & 1 \end{array}\right],$$

     but we cannot proceed further.

Thus we are not able to find A^{-1} in Example 5. since we cannot use row operations on A to obtain I (in the left half of the 3 × 6 matrix). This is easily explained since it is clear that

$$\begin{vmatrix} 1 & 0 & 2 \\ 0 & -1 & -1 \\ 0 & 0 & 0 \end{vmatrix} = 0,$$

which also says that det A = 0, since the elementary row operations do not affect whether a square matrix has a zero determinant or not. (Row operations (1) and (2) will affect a non-zero value--but will never make it zero!) Conversely, if det A = 0, then by using Properties 2, 3 and 5 of Section 2.2, A can be put in a triangular form with a zero on the diagonal (since the product of the diagonal elements must be zero). The diagonalization of A ends there, though, as the zero on the diagonal prohibits any further steps, as illustrated in Example 5. Hence, the matrix A cannot be changed into I using elementary row operations. Therefore, if A^{-1} exists it can always be found using row operations and if A^{-1} does not exist, the row operations will always yield a zero on the diagonal and for this reason the identity cannot be obtained using row operations.

     There are two terms associated with square matrices which should be mentioned before proceeding. If det A = 0, then A is called a singular matrix. Otherwise it is non-singular. The rank for any n × n matrix is the order (size) of the largest square submatrix whose determinant is not zero. (A submatrix of a matrix A is any matrix obtained from A by deleting rows or columns of A).

     This completes our work on matrices and determinants, although certainly there is much more that can be done. The next section will be concerned with the solution of linear systems of equations. In Section 2.1 we saw that n equations in m unknowns could be written in the form

$$AX = B, \qquad (5)$$

where A is an n × m matrix, B is an n × 1 matrix, and X is an m × 1 matrix containing the unknowns to be found. If A is square and non-singular, then

$$X = A^{-1}B \qquad (6)$$

yields the solution to Eq. (5). However, if A is not square, or if A is singular, then Eq. (6) is no longer applicable for finding "solutions", and we must resort to the row operation procedure. Even if Eq. (6) holds, row operations in most applications are easier to use.

To solve the system of equations using row operations, the <u>augmented</u> <u>matrix</u> is introduced. The augmented matrix for the system in Eq. (5) is the matrix A (<u>called the</u> <u>coefficient</u> <u>matrix</u>) with the matrix B added as the right hand column. Thus the augmented matrix in this case is an n × m + 1 matrix. This formulation of the augmented matrix is illustrated in the following example.

Example 6.   For the system

$$x_1 - 2x_2 + 3x_3 = 6$$

$$x_1 - x_2 - x_3 = -4 \qquad (7)$$

$$2x_1 + 3x_2 + 5x_3 = 23$$

the coefficient matrix is

$$A = \begin{bmatrix} 1 & -2 & 3 \\ 1 & -1 & -1 \\ 2 & 3 & 5 \end{bmatrix}$$

while the augmented matrix is

$$\begin{bmatrix} 1 & -2 & 3 & 6 \\ 1 & -1 & -1 & -4 \\ 2 & 3 & 5 & 23 \end{bmatrix}$$

This example also shows the simplifications that are obtained by using the augmented matrix, as we see that the augmented matrix obtained completely describes the system of equations as given by Eqs. (7). In the next lecture we will show how the solution to systems of equations may be found using elementary row operations on the augmented matrix.

Exercises

1. Find the inverse of $\begin{bmatrix} 2 & -3 \\ 1 & 3 \end{bmatrix}$ using the adjoint method and row operation method.

2. Find the formula for the inverse of A $= \begin{bmatrix} a_{11} & a_{12} \\ a_{21} & a_{22} \end{bmatrix}$ using the row operations.

3. Use row operations to find the inverses of each of the following matrices.

a) $\begin{bmatrix} -1 & 2 & -3 \\ 4 & -7 & 9 \\ 2 & -1 & -4 \end{bmatrix}$

b) $\begin{bmatrix} -1 & 2 & 1 \\ 0 & 1 & -2 \\ 1 & -3 & 3 \end{bmatrix}$

c) $\begin{bmatrix} 1 & 3 & 4 \\ 3 & -1 & 4 \\ -1 & 2 & 1 \end{bmatrix}$

d) $\begin{bmatrix} 1 & 0 & 2 & -3 \\ 1 & -1 & 2 & -1 \\ 1 & 2 & 3 & -1 \\ 3 & 4 & 7 & -9 \end{bmatrix}$

Your answers may be checked by calculating AA^{-1}.

4. Find an elementary matrix which performs the indicated row operation on

$$A = \begin{bmatrix} -1 & 2 & -1 & 6 \\ 4 & 0 & 1 & 2 \\ 3 & 5 & 7 & 1 \end{bmatrix}.$$

Then multiply matrices to verify that the desired transformation has been accomplished.

a) Interchange the first and third rows.

b) Add 3 times the first row to the third row.

c) Multiply the second row by two.

5. a) Solve the set of equations

$$x_1 + 2x_2 = 3$$

$$2x_1 + 3x_2 = 1$$

by the formula $X = A^{-1}B$.

b) Once the solution of part a) is found using A^{-1}, is the solution for

$$x_1 + 2x_2 = 4$$

$$2x_1 + 3x_2 = 5$$

as much work as that of part a)?

6. Determine the rank of each of the following matrices.

a) $\begin{bmatrix} 1 & 2 & 5 \\ 0 & -1 & 4 \end{bmatrix}$

b) $\begin{bmatrix} 1 & 3 & -1 \\ 0 & 1 & -2 \\ 0 & 2 & -4 \end{bmatrix}$

c) $\begin{bmatrix} 1 & 4 & -1 & 1 \\ 0 & 2 & 5 & -1 \\ 0 & 0 & 1 & -2 \end{bmatrix}$

d) $\begin{bmatrix} 1 & -1 & -2 & 0 \\ 0 & 0 & 1 & -1 \\ 0 & 0 & 0 & 0 \\ 0 & 0 & 0 & 5 \end{bmatrix}$

7. Using the concept of a singular matrix, determine whether or not each of the following has an inverse.

a) $\begin{bmatrix} 1 & 2 & 5 \\ 0 & -1 & 3 \\ -1 & -1 & 4 \end{bmatrix}$

b) $\begin{bmatrix} 1 & 3 & -1 \\ -2 & -6 & 2 \\ 3 & 4 & 10 \end{bmatrix}$

c) $\begin{bmatrix} 1 & 3 & -2 & 5 \\ 0 & 1 & 2 & 1 \\ 0 & 0 & 3 & 5 \\ 0 & 0 & 3 & 4 \end{bmatrix}$

2.4 SOLUTIONS OF SYSTEMS OF EQUATIONS

This section will basically consist of a series of examples which will illustrate the use of row operations in finding the solutions of systems of linear equations. Several examples are discussed because there are a variety of types of solutions that are possible for arbitrary systems of equations. By a _solution_ of n equations in m unknowns we mean any set of m numbers which satisfy all n equations. What we will find is that there are systems which have no solution, systems that have a unique solution and systems that have more than one solution (an infinite number really).

As we have seen in previous sections, a system of n equations in m unknowns can be written as

$$AX = B, \tag{1}$$

where A is n × m, X is m × 1 and B is n × 1. If B has all zeros as its elements, then Eq. (1) is called _homogeneous_, otherwise it is _nonhomogeneous_. For the homogeneous case there is always the _trivial_ solution (all elements of X are zero); however there may also be other, non-trivial, solutions. The examples deal mainly with the nonhomogeneous systems, as the homogeneous systems are handled in exactly the same way.

The procedure followed on all of the examples shown here is to first find the augmented matrix for the given system of equations and then, by using elementary row operations, find an equivalent matrix that has all zeros below the a_{ii} (i = 1,2,... n) diagonal elements. Such a matrix is said to be in _echelon_ form. It can be shown that equivalent matrices represent systems with identical solutions, therefore, once we have found the echelon matrix we can then find the simplified system it represents and solve that new system for the unknown variables in X. Equations in two unknowns will be discussed in detail first, and then several examples with more unknowns will be shown to illustrate the general case.

Example 1.   Find the solution of the equations:

$$x_1 - x_2 = 1$$

$$\tag{2}$$

$$x_1 + x_2 = 0 .$$

First we find the augmented matrix to be $\begin{bmatrix} 1 & -1 & 1 \\ 1 & 1 & 0 \end{bmatrix}$. Then

the following sequence of row operations are done to determine the equivalent echelon matrix:

$$\begin{bmatrix} 1 & -1 & 1 \\ 1 & 1 & 0 \end{bmatrix} \sim \begin{bmatrix} 1 & -1 & 1 \\ 0 & 2 & -1 \end{bmatrix}$$

$$\sim \begin{bmatrix} 1 & -1 & 1 \\ 0 & 1 & -\frac{1}{2} \end{bmatrix}$$

The echelon matrix just obtained is the augmented matrix for the system

$$x_1 - x_2 = 1$$

$$x_2 = -\frac{1}{2}.$$

The solution to this last system is found by substituting $x_2 = -\frac{1}{2}$ into the first equation, which yields

$$X = \begin{bmatrix} x_1 \\ x_2 \end{bmatrix} = \begin{bmatrix} \frac{1}{2} \\ -\frac{1}{2} \end{bmatrix} \tag{3}$$

as the solution. Since the echelon matrix is equivalent to the original augmented matrix, Eq. (3) must also be the solution of the system (2).

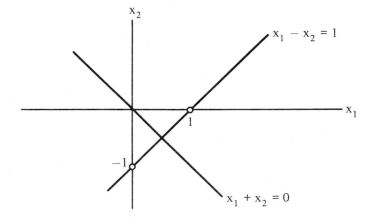

Figure 1.

The solution of Example 1. may be easily verified by substituting into Eqs. (2).

A system of two equations in two unknowns may be interpreted graphically. In Example 1, for instance, the two equations may be taken as equations of straight lines in the x_1, x_2 plane, as shown in Figure 1. In this case the two lines intersect in one point to yield a <u>unique</u> solution. If the two lines were parallel, there would be no solution, as shown in Example 2., or if the lines coincided, there would be an infinite number of solutions as shown in Example 3.

Example 2.   Find, if possible, the solution of

$$x_1 - x_2 = 1$$

$$x_1 - x_2 = 2 \ .$$

(4)

The augmented matrix is

$$\begin{bmatrix} 1 & -1 & 1 \\ 1 & -1 & 2 \end{bmatrix} ,$$

which is equivalent to the matrix

$$\begin{bmatrix} 1 & -1 & 1 \\ 0 & 0 & 1 \end{bmatrix} .$$

This latter matrix is the augmented matrix for the system

$$x_1 - x_2 = 1$$

$$0x_2 = 1 \ ,$$

which yields the inconsistent result that   0   must equal 1. Thus we must conclude that no points satisfy both of the original equations and therefore we say there is no solution.

Example 3.   Find the solution of

$$x_1 - \ x_2 = 1$$

$$2x_1 - 2x_2 = 2 \ .$$

(5)

Proceeding as in the previous examples, we have

$$
\begin{bmatrix} 1 & -1 & 1 \\ 2 & -2 & 2 \end{bmatrix} \sim \begin{bmatrix} 1 & -1 & 1 \\ 0 & 0 & 0 \end{bmatrix}
$$

so that the original Eqs. (5) have the same solutions as

$$x_1 - x_2 = 1$$

$$0x_2 = 0 \quad .$$

(6)

In this case we solve the first equation for x_1 to obtain

$$x_1 = 1 + x_2 .$$

Thus

$$
X = \begin{bmatrix} x_1 \\ x_2 \end{bmatrix} = \begin{bmatrix} 1 + x_2 \\ x_2 \end{bmatrix} = \begin{bmatrix} 1 \\ 0 \end{bmatrix} + x_2 \begin{bmatrix} 1 \\ 1 \end{bmatrix}
$$

(7)

is the "solution" for the original system (5).

Equation (7) really represents an infinite number of solutions to the system (5), as the value of x_2 in Eq. (7) is left unspecified. This situation arises since the original system, which looks like two equations in two unknowns, is really only one equation in the two unknowns, as we see more explicitly in Eqs. (6). The x_2 appearing on the right hand side of Eq. (7) is frequently replaced by a constant such as c.
The three examples shown so far represent the three possibilities that can exist in the solution of systems of equations. In Example 1, for instance, the coefficient matrix has a nonzero determinant and the solution is unique ($X = A^{-1}B$), while in Examples 2. and 3. the coefficient matrices have zero determinants and the existance of a solution depends on the value of the constant matrix B. As the reader may recall, these results very closely parallel the solution of the scalar equation ax = b which fall into the three possibilities:

1) If $a \neq 0$, then x = b/a is the unique solution.

2) If a = 0 and

      a) if $b \neq 0$, then there is no finite solution.

      b) if b = 0, then there is an infinite number of solutions.

The further systems examples to be discussed will also fall into this pattern with det A replacing a, A^{-1} replacing $1/a$ and B replacing b, although the conditions 2a) and 2b) cannot be written as easily.

Before going on to larger systems, two examples of homogeneous equations are shown to complete the analysis for two equations.

Example 4.   Solve the homogeneous system

$$x_1 - x_2 = 0$$

$$x_1 + x_2 = 0 \ . \tag{8}$$

Proceeding as before we have

$$\begin{bmatrix} 1 & -1 & 0 \\ 1 & 1 & 0 \end{bmatrix} \sim \begin{bmatrix} 1 & -1 & 0 \\ 0 & 2 & 0 \end{bmatrix}$$

Thus

$$X = \begin{bmatrix} x_1 \\ x_2 \end{bmatrix} = \begin{bmatrix} 0 \\ 0 \end{bmatrix}$$

is the unique solution, found by solving the system corresponding to the echelon matrix.

Example 5.   Solve the homogeneous system

$$x_1 - x_2 = 0$$

$$2x_1 - 2x_2 = 0 \ . \tag{9}$$

Again we have

$$\begin{bmatrix} 1 & -1 & 0 \\ 2 & -2 & 0 \end{bmatrix} \sim \begin{bmatrix} 1 & -1 & 0 \\ 0 & 0 & 0 \end{bmatrix}$$

and thus

$$X = \begin{bmatrix} x_1 \\ x_2 \end{bmatrix} = \begin{bmatrix} x_2 \\ x_2 \end{bmatrix} = x_2 \begin{bmatrix} 1 \\ 1 \end{bmatrix} \tag{10}$$

yields an infinite set of solutions for Eqs. (9).

Notice that for the homogeneous cases, the straight lines, which the equations represent, must pass through the origin and hence the zero solution must always exist.
In Example 5. the coefficient matrix has a zero determinant and non-zero solutions
exist, while in Example 4, the coefficient matrix has a nonzero determinant and hence
the solution is unique. Equations (7) and (10) should be compared with the analogous
situation in differential equations. That is, for the nonhomogeneous case, say, the

solution as shown in Eq. (7) is composed of a particular solution, $\begin{bmatrix} 1 \\ 0 \end{bmatrix}$, and a

complementary solution, $c \begin{bmatrix} 1 \\ 1 \end{bmatrix}$.

Occasionally the situation arises where three equations in two unknowns must be
considered. Referring to Figure 2, we see that if a third line is added which does
not go through the intersection of the first two, then there will be no solution.
However, if the third line goes through the point of intersection, then the solution
of the first two equations will also satisfy the third equation.

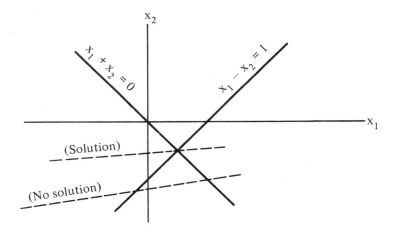

Figure 2.

The above examples and discussion have covered most of the possibilities for
linear equations involving two unknowns. The situation for three or more unknowns
is even more complicated, particularly in the case when the coefficient matrix has
a zero determinant. The general results will be stated here, without proof, so that
they may be verified as we go along:

A system of linear equations AX = B has a solution
if and only if the rank of the augmented matrix is

equal to the rank of the coefficient matrix. The
solution is unique if and only if the rank is the
same as the number of unknowns.

For the homogeneous case the coefficient and augmented matrices always have the
same rank and hence a solution always exists. The student should look at the previous
examples and verify the above results.

The following examples are presented to illustrate the various basic results
for systems of three or more equations. Each example is discussed in the light of
the above general results.

Example 6.   Find the solution of the following system of equations

$$2x_1 - 3x_2 + x_3 = 4$$

$$x_1 + x_2 - x_3 = 2 \tag{11}$$

$$4x_1 - x_2 \quad = 1 \quad .$$

The approach used here is exactly the same as that used on systems of
two equations. The augmented matrix is obtained and then row operations
are used on it to obtain an equivalent matrix in echelon form as shown:

$$
\begin{bmatrix} 2 & -3 & 1 & 4 \\ 1 & 1 & -1 & 2 \\ 4 & -1 & 0 & 1 \end{bmatrix}
\sim
\begin{bmatrix} 1 & 1 & -1 & 2 \\ 2 & -3 & 1 & 4 \\ 4 & -1 & 0 & 1 \end{bmatrix}
$$

$$
\sim
\begin{bmatrix} 1 & 1 & -1 & 2 \\ 0 & -5 & 3 & 0 \\ 0 & -5 & 4 & -7 \end{bmatrix}
$$

$$
\sim
\begin{bmatrix} 1 & 1 & -1 & 2 \\ 0 & -5 & 3 & 0 \\ 0 & 0 & 1 & -7 \end{bmatrix} .
$$

From the equivalent echelon matrix we see that the rank of the coefficient
and augmented matrices are both 3 and hence there is a unique solution.
The solution may be found from the corresponding system of equations:

$$x_1 + x_2 - x_3 = 2$$
$$- 5x_2 + 3x_3 = 0 \qquad (12)$$
$$x_3 = -7$$

or by continuing row operations to get the equivalent matrix

$$\begin{bmatrix} 1 & 0 & 0 & -4/5 \\ 0 & -1 & 0 & 21/5 \\ 0 & 0 & 1 & -7 \end{bmatrix}.$$

Both approaches yield the solution

$$X = \begin{bmatrix} x_1 \\ x_2 \\ x_3 \end{bmatrix} = \begin{bmatrix} -4/5 \\ -21/5 \\ -7 \end{bmatrix} \qquad (13)$$

which may be verified to satisfy all three equations in (11). The
corresponding homogeneous case here has only the zero solution since
the rank of the augmented matrix is equal to the number of unknowns.

In the following examples only the echelon matrices will be considered since
the above examples have amply illustrated the use of elementary row operations that
are used to go from the original augmented matrices to the equivalent echelon matrices.

Example 7.   Find the solution of the system of equations corres-
                      ponding to the echelon matrix:

$$\begin{bmatrix} 1 & 8 & -4 & 6 \\ 0 & -2 & 1 & 0 \\ 0 & 0 & 0 & 1 \end{bmatrix}.$$

If the equations corresponding to this matrix were written out it would
be found that the last row implies 0 = 1. Thus there is no solution.

In the preceding example it can be seen that the coefficient matrix, which would be

$$\begin{bmatrix} 1 & 8 & -4 \\ 0 & -2 & 1 \\ 0 & 0 & 0 \end{bmatrix}$$

and the augmented matrix (the given matrix) do not have the same rank, and thus the fact that there is no solution is consistent with the results stated previously.

Example 8.   For the echelon matrix

$$\begin{bmatrix} 1 & 8 & -4 & 6 \\ 0 & -2 & 1 & 0 \\ 0 & 0 & 0 & 0 \end{bmatrix}$$

we find that the necessary ranks are equal. Thus the solution may be found from:

$$x_1 + 8x_2 - 4x_3 = 6$$

$$- 2x_2 + x_3 = 0 \ .$$

By solving the last equation for x_2 and substituting in the first equation, we find that

$$X = \begin{bmatrix} x_1 \\ x_2 \\ x_3 \end{bmatrix} = \begin{bmatrix} 6 \\ \tfrac{1}{2}x_3 \\ x_3 \end{bmatrix} = \begin{bmatrix} 6 \\ 0 \\ 0 \end{bmatrix} + x_3 \begin{bmatrix} 0 \\ \tfrac{1}{2} \\ 1 \end{bmatrix} \tag{14}$$

For a system of four equations other possibilities exist, as shown in Example 9.

Example 9.   Find the solutions of the system of four equations in four unknowns represented by the augmented matrix:

$$\begin{bmatrix} 1 & 0 & 1 & 6 & 7 \\ 0 & -2 & 1 & 0 & 1 \\ 0 & 0 & 0 & 0 & 0 \\ 0 & 0 & 0 & 0 & 0 \end{bmatrix} .$$

Once again the two necessary ranks are equal and the solution may be found from

$$x_1 \quad + x_3 + 6x_4 = 7$$

$$- 2x_2 + x_3 \qquad = 1 \ . \tag{15}$$

These yield

$$x_1 = 7 - x_3 - 6x_4$$

$$x_2 = -\tfrac{1}{2} + \tfrac{1}{2} x_3$$

which can be written in the form

$$X = \begin{bmatrix} x_1 \\ x_2 \\ x_3 \\ x_4 \end{bmatrix} = \begin{bmatrix} 7 - x_3 - 6x_4 \\ -\tfrac{1}{2} + \tfrac{1}{2} x_3 \\ x_3 \\ x_4 \end{bmatrix}$$

or

$$X = \begin{bmatrix} 7 \\ -\tfrac{1}{2} \\ 0 \\ 0 \end{bmatrix} + x_3 \begin{bmatrix} -1 \\ \tfrac{1}{2} \\ 1 \\ 0 \end{bmatrix} + x_4 \begin{bmatrix} -6 \\ 0 \\ 0 \\ 1 \end{bmatrix} \tag{16}$$

where x_3 and x_4 may be chosen arbitrarily. The form of the answer as given in Eq. (16) may vary since Eq. (15) may be solved for other variables.

The above examples serve to illustrate some of the different situations that can arise in solving systems of equations. The last example, in particular, shows that for larger systems there can be more than one arbitrary constant in the solution. In a more complete discussion of linear equations various criteria can be derived to cover all possible solutions. However, for this discussion, the general result stated previously is sufficient and the examples suffice as an extension to cover the case when an infinite number of solutions exist.

In summary, then, general systems of equations can be solved by the application of elementary row operations on the augmented matrix to find an equivalent echelon matrix. If a solution exists, then it can be found from the echelon form. If a solution does not exist, then at least one row of the echelon form will yield an inconsistent equation such as $0 = 1$. Since the row operations do not affect the ranks of the corresponding matrices, we may then conclude the same results hold for the original system.

Exercises

Find the solutions, if possible, of the following systems of equations, and compare the results to the examples.

1. $\begin{aligned} x_1 + x_2 + x_3 &= 4 \\ x_1 + 2x_2 + 2x_3 &= 2 \\ x_1 + x_2 - x_3 &= -2 \end{aligned}$

2. $\begin{aligned} 4x_1 + 3x_2 - x_3 &= 2 \\ 10x_1 - 6x_2 - 16x_3 &= 5 \\ 6x_1 + 7x_2 + x_3 &= 3 \end{aligned}$

3. $\begin{aligned} x_1 + x_2 + x_3 &= 6 \\ x_1 - 3x_2 - 3x_3 &= -4 \\ 5x_1 - 3x_2 - 3x_3 &= 8 \end{aligned}$

4. $\begin{aligned} x_1 + x_2 + x_3 &= 0 \\ 2x_1 + 3x_2 + x_3 &= 0 \\ 4x_1 - x_2 + 2x_3 &= 0 \end{aligned}$

5. $\begin{aligned} x_1 + 2x_2 + x_3 &= 0 \\ x_1 - 2x_2 - 3x_3 &= 0 \\ 5x_1 + 2x_2 - 3x_3 &= 0 \end{aligned}$

6. $\begin{aligned} x_1 + 2x_2 &= 1 \\ 3x_1 + x_2 &= 8 \\ x_1 - 4x_2 &= 7 \end{aligned}$

7. $\begin{aligned} 2x_1 + x_2 &= 2 \\ x_1 - 4x_2 &= 3 \\ 3x_1 + 2x_2 &= 1 \end{aligned}$

8. $\begin{aligned} x_1 + x_2 + x_3 + x_4 &= 4 \\ x_1 + 3x_2 + 3x_3 &= 2 \\ x_1 + x_2 + 2x_3 - x_4 &= 6 \end{aligned}$

9. $\begin{aligned} x_1 - x_2 - 3x_3 - x_4 &= 1 \\ 2x_1 + 4x_2 + 2x_4 &= -2 \\ 3x_1 + 4x_2 - 2x_3 &= 0 \\ x_1 + 2x_2 - 3x_4 &= 3 \end{aligned}$

10. $\begin{aligned} x_1 - x_2 - 3x_3 - x_4 &= 1 \\ x_1 + 2x_2 - 3x_4 &= 3 \\ 3x_1 - 6x_3 - 5x_4 &= 5 \\ 2x_1 + x_2 - 3x_3 - 4x_4 &= 4 \end{aligned}$

3

PARTIAL DIFFERENTIATION

3.1 PARTIAL DERIVATIVES AND DIFFERENTIALS

There are many instances in science and engineering where a quantity is determined by a number of other quantities. A simple example is the volume of a box

$$V = w \, \ell \, h, \tag{1}$$

where the volume V is the product of the width w, length ℓ and height h. More generally, if z is uniquely determined by values of x and y, then we say z is a function of x and y and write

$$z = f(x,y). \tag{2}$$

In fact, z may be a function of several variables, which might be written

$$z = f(x,y,w,u). \tag{3}$$

We will primarily use functions of the form (2) since they are more easily understood in our 3-dimensional world. In Eqs. (2) and (3), x,y,w and u are called <u>independent</u> variables and z the <u>dependent</u> variable.

A three dimensional <u>rectangular</u> coordinate system is constructed by three mutually perpendicular axes as shown in Figure 1. Here the positive axes have been labelled and form a <u>right-handed</u> coordinate system, since if the index finger of the right hand is directed along the positive x axis, then the second finger and the thumb of the right hand are also directed along the positive y and z axes respectively. A point in 3 dimensions is represented by the triplet (x,y,z), where x, y and z are the coordinates of the point. In general a function of the form in Eq. (2) represents a <u>surface</u> in three dimensional space. For the most part these are very hard to sketch or visualize. However, one important example is the hemisphere of radius a and center at (0,0,0):

$$z = \sqrt{a^2 - (x^2 + y^2)} \, . \tag{4}$$

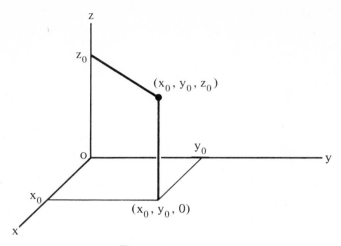

Figure 1.

A sketch of this in the first octant is shown in Figure 2. In general we will not be trying to sketch our surfaces, but we must keep in mind that a function of two variables can be represented by a surface in three space. For functions of more than 2 variables such three dimensional representation is impossible. The terminology hypersurface is often used.

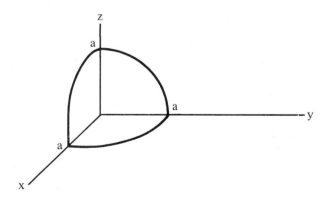

Figure 2.

Example 1.

$$x^2 + y^2 + z^2 + w^2 = a^2$$

(5)

is the equation of a hypersphere.

The ordinary derivative of a function of several variables with respect to one of the independent variables, keeping all other variables constant, is called the partial derivative with respect to the variable. Partial derivatives of $f(x,y)$ with respect to x and y are denoted by $\frac{\partial f}{\partial x}$ and $\frac{\partial f}{\partial y}$ respectively and are given mathematically by:

$$\frac{\partial f}{\partial x} = \lim_{\Delta x \to 0} \frac{f(x+\Delta x, y) - f(x,y)}{\Delta x} \tag{6}$$

$$\frac{\partial f}{\partial y} = \lim_{\Delta y \to 0} \frac{f(x, y+\Delta y) - f(x,y)}{\Delta y} . \tag{7}$$

Example 2.   If

$$f(x,y) = 3xy + 2x \cos xy,$$

then

$$\frac{\partial f}{\partial x} = 3y + 2 \cos xy - 2xy \sin xy,$$

and

$$\frac{\partial f}{\partial y} = 3x - 2x^2 \sin xy.$$

Example 3.   If

$$z = \ln(x^2 + y^2),$$

then

$$\frac{\partial z}{\partial x} = \frac{2x}{x^2 + y^2} \quad \text{and} \quad \frac{\partial z}{\partial y} = \frac{2y}{x^2 + y^2} .$$

For functions of one variable the derivative yields the slope of the curve. For functions of two variables a similar concept is shown in Figure 3, where the plane $y = y_0$ has cut the surface $z = f(x,y)$ in a curve that has the slope $\frac{\partial f}{\partial x}(x_0, y_0)$ at the point (x_0, y_0, z_0). The plane $x = x_0$ similarly cuts the surface in a different curve, which has a slope $\frac{\partial f}{\partial y}(x_0, y_0)$ at the same point. Other notations used for partial derivatives of $f(x,y)$ are:

$$\frac{\partial f}{\partial x} = f_x = f_x(x,y) = \left(\frac{\partial f}{\partial x}\right)_y \tag{8}$$

and similarly for $\frac{\partial f}{\partial y}$. The last notation, $\left(\frac{\partial f}{\partial x}\right)_y$, is used mainly when f is a

function of more variables and the subscripts outside the parenthesis denote the

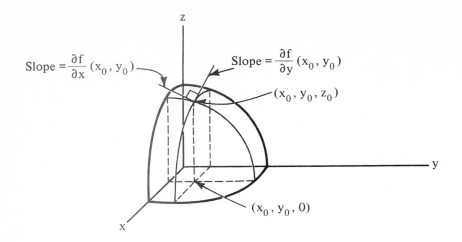

Figure 3.

variables being held constant.

As can be seen from the definition and the above examples, the difference in the partial derivative, $\frac{\partial f}{\partial x}$, and the ordinary derivative, $\frac{df}{dx}$, is largely in the point of view desired. For example $f(x) = e^{\alpha x}$ has the derivative $\frac{df}{dx} = \alpha e^{\alpha x}$ while $f(x,y) = e^{xy}$ has the derivative $\frac{\partial f}{\partial x} = ye^{xy}$. In both cases the differentiation process is the same, but in the latter function we wish to call attention to the presence of two variables.

Higher derivatives are denoted as follows:

$$\frac{\partial^2 f}{\partial x^2} = \frac{\partial}{\partial x}(\frac{\partial f}{\partial x}) \quad , \quad \frac{\partial^2 f}{\partial y^2} = \frac{\partial}{\partial y}(\frac{\partial f}{\partial y}) \tag{9}$$

and

$$\frac{\partial^2 f}{\partial x \partial y} = \frac{\partial}{\partial x}(\frac{\partial f}{\partial y}) \quad , \quad \frac{\partial^2 f}{\partial y \partial x} = \frac{\partial}{\partial y}(\frac{\partial f}{\partial x}) \; . \tag{10}$$

The derivatives denoted in Eqs. (9) are straightforward and do not differ from the situation for ordinary derivatives. The derivatives denoted in Eqs. (10) are called mixed partial derivatives. A very important result is that if the above derivatives are <u>continuous</u>, then

$$\frac{\partial^2 f}{\partial x \partial y} = \frac{\partial^2 f}{\partial y \partial x} \; . \tag{11}$$

Example 4.   If $f(x,y) = x \cos y$,

                    then

$$\frac{\partial f}{\partial x} = \cos y \quad \text{and} \quad \frac{\partial f}{\partial y} = -x \sin y .$$

Thus

$$\frac{\partial^2 f}{\partial y \partial x} = -\sin y,$$

and

$$\frac{\partial^2 f}{\partial x \partial y} = -\sin y.$$

The next important topic to be covered is that of the <u>differential</u> for a function of two variables. This will be introduced using the <u>tangent plane</u> to the surface $z = f(x,y)$. Before doing this let's review the situation for a function of one variable. If $y = f(x)$, then dy is shown geometrically in Figure 4. The analytical

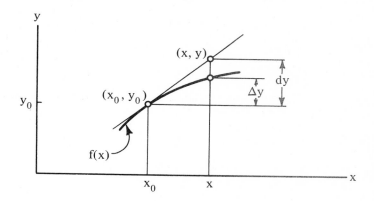

Figure 4.

expression for dy is obtained from the equation of the tangent line. Using Figure 4, we see that the tangent line is given by

$$\frac{y - y_0}{x - x_0} = f'(x_0) \tag{12}$$

or

$$y - y_0 = f'(x_0)(x - x_0), \tag{13}$$

where Eq. (12) is obtained by equating the two expressions for the slope of the line. From Eq. (13), if we let $dy = y - y_0$ and $dx = x - x_0$, we get the <u>differential</u>:

$$dy = f'(x_0)dx . \tag{14}$$

The differential may be used in several ways. We will use it mainly to obtain the chain rule in the next section, but from Figure 4, dy as given in Eq. (14) is an approximation to the quantity

$$\Delta y = f(x) - f(x_0), \tag{15}$$

when $x - x_0$ is "small".

The situation for functions of two variables is very similar. The tangent plane to $z = f(x,y)$ is given by:

$$A (x - x_0) + B (y - y_0) - (z - z_0) = 0 \tag{16}$$

where

$$A = \frac{\partial f}{\partial x} (x_0, y_0), \qquad B = \frac{\partial f}{\partial y} (x_0, y_0). \tag{17}$$

This plane is actually shown in Figure 3. since it is the plane formed by the two tangent lines which have the slope $\frac{\partial f}{\partial x}$ and $\frac{\partial f}{\partial y}$.

Example 5.   Find the tangent plane to

$$z = y^2 e^x$$

at $(0,3)$.

The values of A and B are found as

$$A = \frac{\partial f}{\partial x} (0,3) = y^2 e^x \bigg|_{(0,3)} = 9$$

and

$$B = \frac{\partial f}{\partial y} (0,3) = 2y e^x \bigg|_{(0,3)} = 6.$$

The correct value of z_0 is 9, from the given function, and thus the tangent plane is given by

$$9(x - 0) + 6(y - 3) - (z - 9) = 0,$$

or

$$9x + 6y - z = 9.$$

If we let $dx = x - x_0$, $dy = y - y_0$, and $dz = z - z_0$ in Eq. (16) we obtain the differential for a function of two variables:

$$dz = \frac{\partial f}{\partial x} dx + \frac{\partial f}{\partial y} dy, \tag{18}$$

where it is understood that the derivatives are evaluated at the point (x_0, y_0).
Equation (18) will be used to develop the chain rule in the next section and can
also be used for approximating the quantity

$$\Delta z \;=\; f(x,y) - f(x_0, y_0). \qquad\qquad (19)$$

This approximation is illustrated in Figure 5. If the point (x,y) is moved from

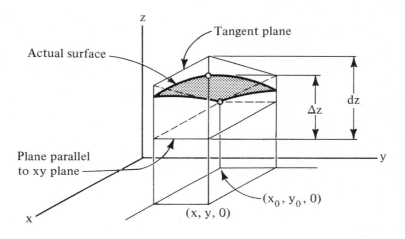

Figure 5.

the point (x_0, y_0) then dz gives the change along the tangent plane surface, while
Δz gives the exact change along the given surface $z = f(x,y)$.

Example 6.   Calculate Δz and dz for $z = x^2 + xy$.
     From Eq. (18) we get

$$dz = (2x_0 + y_0)dx + x_0 dy.$$

Letting $x = x_0 + \Delta x$, $y = y_0 + \Delta y$ in Eq. (19) we have

$$\Delta z = (x_0 + \Delta x)^2 + (x_0 + \Delta x)(y_0 + \Delta y) - (x_0^2 + x_0 y_0)$$

$$= x_0^2 + 2x_0\Delta x + \Delta x^2 + x_0 y_0 + \Delta x y_0 + x_0\Delta y$$

$$+ \Delta x\Delta y - x_0^2 - x_0 y_0$$

$$= (2x_0 + y_0)\Delta x + x_0\Delta y + \Delta x^2 + \Delta x\Delta y.$$

where $\Delta x = x - x_0$, $\Delta y = y - y_0$.

Comparing the expression for dz and Δz in Example 6, we see that dz does
represent the major portion of Δz and hence if Δx and Δy are small dz will

approximate Δz. The advantage is that in general dz is much easier to calculate than Δz.

Exercises

1. Let $W = \sqrt{2 - r^2 - 2s^2}$

    a) What are the independent variables?
    b) What is the dependent variable?
    c) For what values of r and s is W a function of r and s?
    d) What possible values of W are there?
    e) Sketch the surface in the first octant using a right-handed
        r,s, W rectangular coordinate system.

2. For each of the following functions find all possible values of x and y
    for which the functions are defined. In each case what possible values
    of z are there?

    a) $z = \dfrac{4}{x^2 + y^2 - 9}$ b) $z = \dfrac{2}{x^2 + y^2 + 1}$

    c) $z = \ln(4 - \sqrt{x^2 + y^2 - 1})$ d) $z = 4 - e^{-xy}$

3. Find $\dfrac{\partial f}{\partial x}$ and $\dfrac{\partial f}{\partial y}$ for each of the following functions.

    a) $f(x,y) = e^{-x} \cos y$

    b) $f(x,y) = \tan^{-1}(y/x)$ $\left(\dfrac{d}{du} \tan^{-1} u = \dfrac{1}{1+u^2}\right)$

    c) $f(x,y) = \ln \sqrt{x^2 + y^2}$

4. Find f_{xx}, f_{yy}, f_{xy} and f_{yx} for $f(x,y) = x^2 \cos y$ and compare the mixed partial
    derivatives.

5. Find the tangent plane to each of the following surfaces at the indicated points.

    a) $z = x^2 + 2xy$ at $(1,1)$ b) $z = 3yx^2 + 2y^3$ at $(1,2)$

    c) $W = r^2 \cos \theta$ at $(r,\theta) = (2,\pi)$ d) $z = x^2 e^{xy}$ at $(1,2)$

6. Use Eq. (18) to approximate $(1.1)^2 + (1.9)^2$ and compare the approximation to the exact value.

    Hint: let $z = x^2 + y^2$, $x_0 = 1$, $y_0 = 2$.

7. Find an approximate value for $z = \ln(x^2 + 2y^2)$ at the point $(1.1, .2)$.

8. Suppose it is known that a certain function $z = f(x,y)$ has the experimental value of $f(2,1) = 5$ and the experimental slopes $f_x(2,1) = 2$ and $f_y(2,1) = 1.5$. Find estimates for values of $f(1.9, 1.2)$, $f(1.9, 1.4)$ and $f(1.9, .8)$.

3.2   THE CHAIN RULE, JACOBIANS AND IMPLICIT DIFFERENTIATION

The concept of the <u>chain rule</u> is very important in many applications. This will be reviewed first for functions of one variable and then discussed in various forms for functions of several variables. In the last section Eq. (14) gave the differential for a function of one variable:

$$dy \; = \; f'(x) \, dx \; = \; \frac{df}{dx} \, dx \tag{1}$$

where the subscript on x_0 has been dropped for convenience. The chain rule is used when y is a function of x and x is a function of another variable, say t. That is:

$$y \; = \; f(x), \qquad x = x(t). \tag{2}$$

Since x is a function of t, then y will also be a function of t and hence it is possible to calculate $\frac{dy}{dt}$. The derivative $\frac{dy}{dt}$ may be found by "dividing" both sides of Eq. (1) by dt, yielding

$$\frac{dy}{dt} \; = \; \frac{df}{dx} \, \frac{dx}{dt} \; , \tag{3}$$

which is the chain rule for functions of one variable. The "dividing" by dt is not mathematically rigorous, but will suffice for our purposes. As the following examples show, we frequently use the chain rule without knowing it.

Example 1.   If

$$y = \cos x \text{ and } x = 3t^2$$

then
$$\frac{dy}{dt} \; = \; (- \sin x) \, 6t.$$

Writing this in terms of one variable, we have

$$\frac{dy}{dt} = -6t \sin 3t^2 .$$

The final form of the derivative in Example 1. is recognized as the derivative of $\cos 3t^2$ even when the chain rule is not used explicitly.

Example 2.   Find $\frac{dy}{dx}$ if y is of the form

$$y = f(x^2).$$

To find $\frac{dy}{dx}$, let $z = x^2$, then

$$y = f(z)$$

from which we obtain

$$\frac{dy}{dx} = \frac{df}{dz} \frac{dz}{dx} = \frac{df}{dz} 2x .$$

Again the reader may recognize this general result when it is applied to a familiar function such as $y = e^{x^2}$.

The chain rule for functions of two or more variables takes on several forms which can be obtained from the differential derived in the preceding section:

$$dz = \frac{\partial f}{\partial x} dx + \frac{\partial f}{\partial y} dy. \qquad (4)$$

Now, if $z = f(x,y)$, $x = x(t)$ and $y = y(t)$, we have in the same fashion as for functions of one variable

$$\frac{dz}{dt} = \frac{\partial f}{\partial x} \frac{dx}{dt} + \frac{\partial f}{\partial y} \frac{dy}{dt} . \qquad (5)$$

If $z = f(x,y)$, $x = x(r,s)$ and $y = y(r,s)$ then z is also a function of r and s. Hence

$$\frac{\partial z}{\partial r} = \frac{\partial f}{\partial x} \frac{\partial x}{\partial r} + \frac{\partial f}{\partial y} \frac{\partial y}{\partial r} \qquad (6)$$

where the change from $\frac{dz}{dr}$ to $\frac{\partial z}{\partial r}$ is made to indicate that z is a function of two variables. The change in notation is permissible since, as we saw in the last section, partial differentiation and ordinary differentiation are exactly the same mathematical processes. Of course there is an equation analogous to Eq. (6) for $\frac{\partial z}{\partial s}$. The final form for the chain rule considers the case $z = f(x)$ and $\dot{x} = x(r,s)$. In this case Eq. (4) will yield

$$\frac{\partial z}{\partial s} = \frac{df}{dx} \frac{\partial x}{\partial s} \ . \tag{7}$$

Equations (5), (6) and (7) represent the three different forms of the chain rule for functions of two variables. Equations (5) and (6) are easily extended to functions of more variables by adding the appropriate terms to the right sides. The following examples will illustrate the use of the chain rule.

Example 3.   Find $\frac{dz}{dt}$ if

$$z = \sin (e^x + y)$$

and

$$x = t^2 + 1, \quad y = e^t.$$

Using Eq. (5) we obtain

$$\frac{dz}{dt} = [\cos(e^x + y) \ e^x] \ 2t + [\cos(e^x + y)]e^t.$$

Example 4.   Find $\frac{\partial z}{\partial r}$ and $\frac{\partial z}{\partial \theta}$ if

$$z = x^2 + y^2$$

where

$$x = e^r \cos \theta \text{ and } y = e^r \sin \theta.$$

Using Eq. (6) we obtain

$$\frac{\partial z}{\partial r} = 2x \ e^r \cos \theta + 2y \ e^r \sin \theta$$

and

$$\frac{\partial z}{\partial \theta} = 2x \ (-e^r \sin \theta) + 2y \ e^r \cos \theta.$$

In Example 3. the chain rule has been used more than once since it is used when taking the derivative of the sine function and then again to find $\frac{dx}{dt}$ and $\frac{dy}{dt}$. In both examples the format of the final answer could be alternated by substituting for x and y. How far this type of process is continued depends on the particular problem under consideration.

Example 5. Show that $\frac{\partial w}{\partial x} - \alpha \frac{\partial w}{\partial y} = 0$ is reduced to $\frac{\partial w}{\partial r} = 0$

if $r = y - \alpha x$, $s = y + \alpha x$ and α is a nonzero constant.

If

$$w = F(r,s) \text{ then}$$

$$\frac{\partial w}{\partial x} = \frac{\partial w}{\partial r} \frac{\partial r}{\partial x} + \frac{\partial w}{\partial s} \frac{\partial s}{\partial x} = -\alpha \frac{\partial w}{\partial r} + \alpha \frac{\partial w}{\partial s}$$

and

$$\frac{\partial w}{\partial y} = \frac{\partial w}{\partial r} \frac{\partial r}{\partial y} + \frac{\partial w}{\partial s} \frac{\partial s}{\partial y} = \frac{\partial w}{\partial r} + \frac{\partial w}{\partial s} ,$$

where the given equations for r and s are used to find the necessary derivatives of r and s. Thus

$$\frac{\partial w}{\partial x} - \alpha \frac{\partial w}{\partial y} = -2\alpha \frac{\partial w}{\partial r} = 0,$$

which gives the desired result since $\alpha \neq 0$.

Example 5. illustrates the use of the chain rule as it applies to a broad class of problems that arise in many different areas of applications. The next example shows how the chain rule can actually be used in successive applications to the same problem.

Example 6. Suppose that

$$z = f(x,y)$$

where

$$x = x(r,s) \text{ and } y = y(r,s),$$

and

$$r = r(t) \text{ and } s = s(t).$$

In this case, as in previous examples, z will be a function of t when r and s are substituted into x and y, which in turn are substituted into z. Thus

$$\frac{dz}{dt} = \frac{\partial f}{\partial x} \frac{dx}{dt} + \frac{\partial f}{\partial y} \frac{dy}{dt}$$

from Eq. (5). Now Eq. (5) can be used again to find $\frac{dx}{dt}$ and $\frac{dy}{dt}$, so that:

$$\frac{dz}{dt} = \frac{\partial f}{\partial x} \left[\frac{\partial x}{\partial r} \frac{dr}{dt} + \frac{\partial x}{\partial s} \frac{ds}{dt} \right] + \frac{\partial f}{\partial y} \left[\frac{\partial y}{\partial r} \frac{dr}{dt} + \frac{\partial y}{\partial s} \frac{ds}{dt} \right].$$

Example 7.   Suppose that,

$$v = v(x,y,z,t),$$

where

$$x = x(t), \ y = y(t) \text{ and } z = z(t).$$

Then, the extension of Eq. (5) to more variables gives

$$\frac{dv}{dt} = \frac{\partial v}{\partial x} \frac{dx}{dt} + \frac{\partial v}{\partial y} \frac{dy}{dt} + \frac{\partial v}{\partial z} \frac{dz}{dt} + \frac{\partial v}{\partial t}.$$

Note that the last term should have a $\frac{dt}{dt}$ multiplying it, but this is obviously 1, and thus it is not written.

The last example is particularly important in applications involving moving coordinate systems. In this case $\frac{\partial v}{\partial t}$ represents the rate of change of v as observed at a fixed point while $\frac{dv}{dt}$ is the rate of change of v as observed from the moving coordinate system. The latter is often called the material derivative.

The concept of the Jacobian will be needed when multiple integrals are discussed. Jacobians arise in a variety of ways when dealing with equations involving functions of several variables. To help with the concepts needed for use in multiple integrals, consider the following example.

Example 8.   A region in the xy plane is bounded by the lines x + y = 8, x - 2y = 2 and y = 0. Find the corresponding region in the uv plane if x = u + 2v and y = v - u. Compare the ratio of the areas to the value of the determinant

$$J = \begin{vmatrix} \dfrac{\partial x}{\partial u} & \dfrac{\partial x}{\partial v} \\[2mm] \dfrac{\partial y}{\partial u} & \dfrac{\partial y}{\partial v} \end{vmatrix}.$$

The area in the xy plane is illustrated in Figure 1. Substituting
x = u + 2v and y = v – u into x + y = 8, x – 2y = 2 and y = 0, we
obtain v = 8/3, u = 2/3 and v = u respectively. Thus the bounding
lines of Figure 1. go into straight lines in the uv plane. The
desired region is the triangular region shown in Figure 2. The area

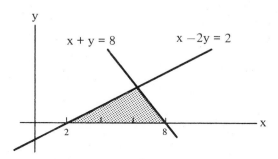

Figure 1.

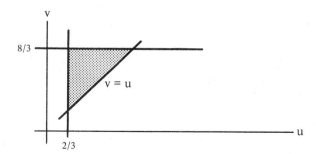

Figure 2.

of the triangle of Figure 1. is 6, while the area of the triangle of
Figure 2. is 2. Finally

$$
J \;=\; \begin{vmatrix} \dfrac{\partial x}{\partial u} & \dfrac{\partial x}{\partial v} \\[2mm] \dfrac{\partial y}{\partial u} & \dfrac{\partial y}{\partial v} \end{vmatrix} \;=\; \begin{vmatrix} 1 & 2 \\ -1 & 1 \end{vmatrix} \;=\; 3
$$

and thus the ratio of the areas is exactly the same as the value of
the determinant.

The determinant used in Example 8. is called the Jacobian of the change of variables $x = u + 2y$ and $y = v - u$.

In general, if the change of variables $x = x(u,v)$ and $y = y(u,v)$ is made, then the Jacobian is defined as in the last example:

$$J = \begin{vmatrix} \dfrac{\partial x}{\partial u} & \dfrac{\partial x}{\partial v} \\[2mm] \dfrac{\partial y}{\partial u} & \dfrac{\partial y}{\partial v} \end{vmatrix} = \frac{\partial(x,y)}{\partial(u,v)} \tag{8}$$

The absolute value of J can be interpreted as the limit of the ratio of an area in the xy plane to the corresponding area in the uv plane as the latter area shrinks to zero. The absolute value is needed since the order in which u and v are chosen will affect the sign of J. As illustrated in the exercises, J can be a function of u and/or v and hence the ratio will not always be a constant, as in Example 8. For a change of variables in three dimensions the Jacobian is defined as

$$J = \begin{vmatrix} \dfrac{\partial x}{\partial u} & \dfrac{\partial x}{\partial v} & \dfrac{\partial x}{\partial w} \\[2mm] \dfrac{\partial y}{\partial u} & \dfrac{\partial y}{\partial v} & \dfrac{\partial y}{\partial w} \\[2mm] \dfrac{\partial z}{\partial u} & \dfrac{\partial z}{\partial v} & \dfrac{\partial z}{\partial w} \end{vmatrix} = \frac{\partial(x,y,z)}{\partial(u,v,w)}, \tag{9}$$

where $x = x(u,v,w)$, $y = y(u,v,w)$ and $z = z(u,v,w)$. The change of variables is also frequently given as $u = u(x,y,z)$, $v = v(x,y,z)$ and $w = w(x,y,z)$. In this case it may be difficult to solve for x, y and z in terms of u,v, and w, but the value of J may still be computed using

$$J = \frac{\partial(x,y,z)}{\partial(u,v,w)} = \frac{1}{\dfrac{\partial(u,v,w)}{\partial(x,y,z)}}. \tag{10}$$

In the last section we saw that the equation

$$x^2 + y^2 + z^2 = a^2 \tag{11}$$

represented a sphere in three dimensional space. Equation (11) is not a function, and yet we may still calculate partial derivatives from it by _implicit_ _differentiation_. Implicit differentiation means that one thinks of Eq. (11) as defining z as a function of x and y:

$$x^2 + y^2 + [z(x,y)]^2 = a^2 \tag{12}$$

and then take the necessary partial derivatives. For Eq. (12) we then have

$$2x + 2[z(x,y)] \frac{\partial z}{\partial x} = 0$$

and

$$2y + 2[z(x,y)] \frac{\partial z}{\partial y} = 0$$

which may be solved for $\frac{\partial z}{\partial x} = -\frac{x}{z}$ and $\frac{\partial z}{\partial y} = -\frac{y}{z}$ respectively. Equation (11)

may be "solved" for z and the above derivatives checked. In general, however, such equations cannot be solved and implicit differentiation is needed.

Example 9.   Find $\frac{\partial z}{\partial x}$ for

$$xy \sin z + z^3 = 1. \tag{13}$$

Rewriting Eq. (13) in the form of Eq. (12) we have

$$xy \sin z(x,y) + z^3(x,y) = 1.$$

Implicit differentiation then gives us

$$y \sin z + xy \cos z \frac{\partial z}{\partial x} + 3 z^2 \frac{\partial z}{\partial x} = 0,$$

which can be solved for $\frac{\partial z}{\partial x}$ to give

$$\frac{\partial z}{\partial x} = \frac{-y \sin z}{xy \cos z + 3z^2}$$

Exercise 7. will illustrate how the concept of implicit differentiation can be extended when several equations are involved.

A word of caution should be mentioned concerning the use of implicit differentiation. Some attempt should be made to justify the assumption that a given equation in x,y and z does define a function z(x,y). For instance one can easily calculate $\frac{\partial z}{\partial x}$ from

$$x^2 + y^2 + z^2 + a^2 = 0, \tag{14}$$

but the derivative is meaningless since there is no function z(x,y) defined by Eq. (14) because of the sum of squares appearing on the left side of the equation. Thus some analysis or computations need to be carried out to verify the assumption that a given equation does define a function.

Exercises

1. If $z = e^{xy^2}$, $x = t \cos t$, $y = t \sin t$, find $\dfrac{dz}{dt}$ at $t = \dfrac{\pi}{2}$.

2. If $z = \sin \dfrac{y}{x}$, $x = 3r^2 + 2s$, $y = 4r - 2s^2$, find $\dfrac{\partial z}{\partial r}$ and $\dfrac{\partial z}{\partial s}$ at $r = 1$ and $s = 0$.

3. If $z = e^x \cos y$ and x and y are defined implicitly by the equations

$$x^2 + 2e^x - t^2 + 3t = 2, \qquad y^2t + t - y = 0,$$

then find $\dfrac{dz}{dt}$ for $t = 0$. (Note that $x = 0$ satisfies the first equation when $t = 0$.)

4. If $y = uv$ where u and v are functions of x, find $\dfrac{dy}{dx}$ by applying Eq. (5) appropriately.

5. If a and b are constants and $w = f(ax + by)$, show that $a\,\dfrac{\partial w}{\partial y} = b\,\dfrac{\partial w}{\partial y}$.

6. Apply Eq. (7) twice to show that $u = f(x + ct)$ satisfies the equation

$$\frac{\partial^2 u}{\partial x^2} = \frac{1}{c^2}\,\frac{\partial^2 u}{\partial t^2}$$

where c is a constant.

7. Polar coordinates $x = r \cos \theta$ and $y = r \sin \theta$ are substituted in a function $w = f(x,y)$.

   a) Show that $\dfrac{\partial w}{\partial r} = f_x \cos \theta + f_y \sin \theta$ and find $\dfrac{\partial w}{\partial \theta}$.

   b) Use part a) to show that

$$\left(\frac{\partial w}{\partial r}\right)^2 + \frac{1}{r^2}\left(\frac{\partial w}{\partial \theta}\right)^2 = f_x^2 + f_y^2.$$

   c) Use part a) to calculate $\dfrac{1}{r}\dfrac{\partial}{\partial r}\left(r\dfrac{\partial w}{\partial r}\right) + \dfrac{1}{r^2}\dfrac{\partial^2 w}{\partial \theta^2}$.

8. A region in the xy plane is bounded by the four lines x + 2y = 2, x + 2y = 4, x − y = 1 and x − y = −2.

   a) Sketch the given region in the xy plane and the corresponding region

      in the uv plane when x $= \dfrac{u + 2v}{3}$ and y $= \dfrac{u - v}{3}$.

   b) Calculate the Jacobian $\dfrac{\partial(x,y)}{\partial(u,v)}$ and verify that it does give the

      stated ratio property.

9. A region in the xy plane is bounded by $x^2 + y^2 = a^2$, $x^2 + y^2 = b^2$
   (0 < a < b), x > 0 and y > 0.

   a) Sketch the corresponding region in the rθ plane if x = r cos θ and
      y = r sin θ (r > 0).

   b) Calculate the Jacobian $\dfrac{\partial(x,y)}{\partial(r,\theta)}$.

10. Calculate $\dfrac{\partial y}{\partial u}$ and $\dfrac{\partial y}{\partial x}$ for the function y(x,u) defined by each of the
    following equations.

    a) ln (uy) + y ln u = x

    b) $xyu + \sin y + 2x^2u^2 = 3$

11. The following two equations in four variables may be thought of as being
    solvable for u and v in terms of x and y. That is, u(x,y) and v(x,y), are
    defined by the given equations. With this in mind find $\dfrac{\partial u}{\partial y}$ and $\dfrac{\partial v}{\partial y}$.

    $$u^2 - v^2 + 3xu - 2y = 0$$

    $$uv + 3xy - 4u + x \sin v = 0$$

3.3 DIFFERENTIATION OF INTEGRALS AND OPTIMIZATION PROBLEMS

One application of the chain rule discussed in the last section is that of the
differentiation of an integral with respect to a variable that appears in the limits
or under the integral sign. The general case is shown in Eq. (1):

$$F(x) = \int_{u_1(x)}^{u_2(x)} f(x,t)dt. \tag{1}$$

Before applying the chain rule to find $\frac{dF}{dx}$ we should recall the Fundamental Theorem of calculus:

$$\frac{d}{dx} \int_a^x f(t)dt = f(x), \tag{2}$$

which relates the derivative and the integral operations. Equation (2) can be generalized to:

$$\frac{d}{dx} \int_a^{u(x)} f(t)dt = f(u(x)) \frac{du}{dx}, \tag{3}$$

in which case the upper limit is any differentiable function of x. The following two examples illustrate the use of Eqs. (2) and (3).

Example 1.   Find the derivative of

$$f(x) = \int_0^{x^2} e^{-t}dt.$$

Using Eq. (3) with $u(x) = x^2$ and $f(t) = e^{-t}$ we find that

$$\frac{df}{dx} = e^{-(x^2)} 2x.$$

Thus

$$\frac{d}{dx} \int_0^{x^2} e^{-t}dt = 2xe^{-x^2}.$$

Example 2.   Find the derivative of

$$f(x) = \int_x^0 t \ln(t)dt,$$

In order to apply Eq. (2) we must interchange the limits of integration:

$$\int_{x}^{0} t \ln(t)dt = - \int_{0}^{x} t \ln(t)dt.$$

Thus Eq. (2) yields

$$\frac{d}{dx} \int_{x}^{0} t \ln(t)dt = - \frac{d}{dx} \int_{0}^{x} t \ln(t)dt$$

$$= - x \ln(x).$$

Returning now to Eq. (1), we rewrite $F(x)$ in the form:

$$F(x) = G(u_1,u_2,x) = \int_{u_1}^{u_2} f(x,\alpha)d\alpha , \qquad (4)$$

since F is explicitly a function of u_1,u_2 and x (under the integral sign). Thus by the chain rule:

$$\frac{dF}{dx} = \frac{\partial G}{\partial u_1} \frac{du_1}{dx} + \frac{\partial G}{\partial u_2} \frac{du_2}{dx} + \frac{\partial G}{\partial x} \frac{dx}{dx} \qquad (5)$$

$$= - f(x,u_1) \frac{du_1}{dx} + f(x,u_2) \frac{du_2}{dx} + \int_{u_1}^{u_2} \frac{\partial f}{\partial x} (x,\alpha)d\alpha, \qquad (6)$$

where we have used Eq. (3) and the result of Example 2. to find the first two terms. The third term follows easily if it is assumed that the integration and differentiation can be interchanged. The result shown in Eq. (6) is called Leibnitz's rule and is most easily remembered when $G(u_1,u_2,x)$ is introduced as in Eq. (4).

Example 3.   The derivative of the function defined by

$$F(x) = \int_{0}^{x} \sin(x-t)dt$$

is given by

$$\frac{dF(x)}{dx} = \int_{0}^{x} \cos(x-t) \, dt + \sin(x-x) \frac{dx}{dx}$$

$$= \int_0^x \cos(x-t)dt.$$

when Eq. (6) is used.

Example 4.   The derivative of the function defined by

$$F(x) = \int_x^{x^2} \frac{\sin xt}{t} dt$$

is given by

$$\frac{dF}{dx} = \int_x^{x^2} t \frac{\cos xt}{t} dt + \frac{\sin(x \cdot x^2)}{x^2} 2x - \frac{\sin(x \cdot x)}{x}$$

or

$$\frac{dF}{dx} = \int_x^{x^2} \cos xt \, dt + \frac{2 \sin x^3}{x} - \frac{\sin x^2}{x}.$$

This last example illustrates a very powerful use of Leibnitz's rule since the function F(x) cannot be found explicitly as it is defined. However, the derivative can still be found, which is important and useful in many applications.

The last topic to be covered under partial differentiation is that of finding relative maximum or minimum values of a function of several variables. If there is a point (a,b) such that

$$f(x,y) > f(a,b) \tag{7}$$

for all points (x,y) sufficiently close to (a,b), then f is said to have a <u>local</u> or <u>relative</u> <u>minimum</u> at (a,b). If the inequality of Eq. (7) is reversed, then f has a local or <u>relative</u> <u>maximum</u> at (a,b). The process of finding relative maximum or minimum points are frequently called <u>optimization</u> <u>problems</u>. A <u>necessary</u> condition that must be satisfied at a relative maximum or minimum is that:

$$\frac{\partial f}{\partial x} = 0 \quad \underline{and} \quad \frac{\partial f}{\partial y} = 0 \quad at \quad (a,b). \tag{8}$$

The necessity of Eq. (8) follows from the geometry of the surface and similar results for functions of one variable. Equations (8) are not a sufficient condition for a relative maximum or minimum to exist though, as the next example shows.

Example 5.   If $f(x,y)$ is defined by

$$f(x,y) = x^2 - y^2$$

then

$$\frac{\partial f}{\partial x} = 2x = 0 \quad \text{and} \quad \frac{\partial f}{\partial y} = -2y = 0.$$

Thus $(0,0)$ is the only possible relative maximum or minimum point. But
$f(0,0) = 0$ is not a relative maximum or minimum value since $f(0,\delta) = -\delta^2 < 0$
no matter how small δ is made, and $f(\varepsilon,0) = \varepsilon^2 > 0$ no matter how small
ε is made. Thus Eq. (7) cannot be satisfied no matter which way the inequality is written.

The point $(0,0)$ of Example 5. is called a saddle point, since the surface in three
dimensional space, which represents the function, resembles a saddle. A saddle
point will occur whenever a function $f(x,y)$ has a maximum behavior in one direction,
the y direction for Example 5, while in another direction it has a minimum behavior.

Sufficient conditions for a relative maximum or minimum are given by the following second derivative test:

If (a,b) is a point (called a <u>critical</u> <u>point</u>) which satisfies Eqs. (8) and if
Δ is given by

$$\Delta = \left\{ f_{xx} f_{yy} - f_{xy}{}^2 \right\} \Bigg|_{(a,b)} \tag{9}$$

then

  1) (a,b) is a relative maximum point if $\Delta > 0$ and $f_{xx} < 0$;

  2) (a,b) is a relative mimimum point if $\Delta > 0$ and $f_{xx} > 0$;

  3) (a,b) is a saddle point if $\Delta < 0$;

  4) No information is obtained if $\Delta = 0$.

The derivation of the second derivative test involves the directional derivative,
which is not discussed until Chapter 5. A comparison to the second derivative test
for functions of one variable will at least show that the second half of the conditions 1) and 2) are consistent with previous results. Note that condition 1) and
2) require that Δ be positive and therefore we must conclude that f_{yy} has the same
sign as f_{xx} in cases 1) and 2). The following examples illustrate the use of Eqs. (8)
and (9).

Example 6.   Find the critical points for the function

$$f(x,y) = x^3 + y^3 - 3x - 12y + 10$$

and then classify them according to the second derivative test.

From Eqs. (8) we obtain the two equations

$$f_x = 3x^2 - 3 = 0$$

$$f_y = 3y^2 - 12 = 0,$$

which have the solutions $x = \pm 1$ and $y = \pm 2$ respectively. Thus there are four critical points, denoted as

$$(a,b) = (\pm 1, \pm 2).$$

Calculating the three second partial derivatives we find that

$$\Delta = 36xy$$

so that

(-1,2) and (1,-2) are saddle points,

(1,2) is a relative minimum point with $f(1,2) = -8$,

and

(-1,-2) is a relative maximum point with $f(-1,-2) = 28$.

Example 7.   What are the dimensions of a rectangular box, without top, if it is to have a minimum surface area and yet hold 32 cubic feet.

If we let x,y and z be the three dimensions of the box then the volume and the surface area are given by:

$$V = xyz = 32$$

and

$$S = xy + 2yz + 2xz.$$

Solving the first equation for $z = \dfrac{32}{xy}$ and substituting into the second we get:

$$S(x,y) = xy + \frac{64}{x} + \frac{64}{y}.$$

Thus

$$\frac{\partial S}{\partial x} = y - \frac{64}{x^2} = 0 \quad \text{or} \quad x^2 y = 64$$

and

$$\frac{\partial S}{\partial y} = x - \frac{64}{y^2} = 0 \quad \text{or} \quad xy^2 = 64.$$

Dividing $x^2 y = 64$ by $xy^2 = 64$ we find that $\frac{x}{y} = 1$. Solving for y and substituting into either equation gives

$$x^3 = 64$$

and hence

$$x = y = 4 \quad \text{and} \quad z = 2$$

is the only critical point. The second derivative test yields

$$\Delta = \left(\frac{128}{x^3}\right)\left(\frac{128}{y^3}\right) - 1 > 0 \quad \text{at} \quad (x,y) = (4,4)$$

and

$$\frac{\partial^2 S}{\partial x^2} = \left(\frac{128}{x^3}\right) > 0 \quad \text{at} \quad (x,y) = (4,4).$$

Thus the desired minimum surface area is given when the dimensions of the box are $4 \times 4 \times 2$.

The above examples illustrate the method of maximinizing or minimizing a function of two variables. For functions of three or more variables, Eqs. (8) are easily extended, but the second derivative test is no longer applicable. In this case physical arguments or the actual evaluation of the function near a critical point will have to be used.

Example 7. is an example of a _constrained_ optimization problem, since we want to minimize the surface area, subject to the constraint that the volume be held constant. Another method of solving such problems is through the use of _Lagrange multipliers_, which is a very powerful tool in many optimization problems. One statement of the use of Lagrange multipliers will be given, but the technique is the same if more variables or more constraints are given, although the equations will differ by the addition of more multipliers.

If $f(x,y,z)$ is to be minimized or maximized subject to the constraint that $g(x,y,z) = 0$, <u>then the method of Lagrange multipliers says that the points where f</u> <u>has a maximum or a minimum are included in the set of simultaneous solutions (x,y,z,λ)</u> <u>of the equations</u>

$$f_x + \lambda g_x = 0$$

$$f_y + \lambda g_y = 0$$

$$f_z + \lambda g_z = 0 \qquad (10)$$

$$g = 0 \ .$$

You will notice that all four equations of (10) are obtained from applying the extension of Eqs. (8) to the function $F(x,y,z,\lambda) = f(x,y,z) + \lambda g(x,y,z)$. If $g(x,y,z) = 0$ at all points, then f has a maximum or minimum whenever F does. Extraneous solutions may appear because of the introduction of λ. The following examples will illustrate the use of Lagrange multipliers.

Example 8.   Find the minimum of

$$S = xy + 2yz + 2xz$$

subject to the constraint

$$xyz - 32 = 0.$$

Let

$$F = xy + 2yz + 2xz + \lambda(xyz - 32)$$

and apply Eqs. (8) or (10):

$$y + 2z + \lambda yz = 0 \qquad (a)$$
$$x + 2z + \lambda xz = 0 \qquad (b)$$
$$2y + 2x + \lambda xy = 0 \qquad (c)$$
$$xyz - 32 = 0. \qquad (d)$$

The simultaneous solution of a) and b) yield

$$z(y - x) = 0 \quad \text{so that} \quad z = 0 \text{ or } x = y$$

and c) and d) yield

$$x(2z - y) = 0 \quad \text{so that} \quad x = 0 \text{ or } y = 2z.$$

Neither $x = 0$ or $z = 0$ will satisfy the constraint and hence $x = y$ and $z = y/2$ are substituted into (d) to give

$$\frac{y^3}{2} = 32 \quad \text{or} \quad y = 4.$$

Thus $x = 4$, $y = 4$ and $z = 2$ is the desired optimum point.

Here the solution must be checked for a maximum or a minimum. Also one must consider whether there are other possible solutions of Eqs. (8). One advantage of Lagrange multipliers is shown here: the constraint does not have to be solved for one of the variables in terms of the others.

Example 9.   Find the shortest distance from (1,0) to the parabola $y^2 = 4x$.

Since the distance is a positive function, it will be minimized whenever the square of the distance is minimized. Thus we wish to minimize

$$f(x,y) = (x - 1)^2 + y^2$$

subject to the constraint that

$$y^2 - 4x = 0.$$

Solution (1): Substitute $y^2 = 4x$ into f to find

$$g(x) = (x - 1)^2 + 4x$$

so that

$$\frac{dg}{dx} = 2(x - 1) + 4 = 2x + 2 = 0$$

which says $x = -1$ is the critical point. But this does not satisfy the constraint $y^2 = 4x$.

Solution (2): Let $F(x,y) = (x - 1)^2 + y^2 + \lambda(y^2 - 4x)$ and apply Eqs. (8) or (10):

$$2(x - 1) - 4\lambda = 0$$
$$2y + 2\lambda y = 0$$
$$y^2 - 4x = 0.$$

The second equation says y = 0 or λ = -1. Substituting λ = -1 in the
first equation gives the previous result, while y = 0 substituted in
the third equation gives x = 0. Thus the shortest distance from (1,0)
to the parabola is 1, and occurs at the point (0,0).

What happened in Example 9. is that the minimum occurred at an "end" point and the
derivative set equal to zero no longer yielded the correct values in the first solution. However, the introduction of the Lagrange multiplier λ avoids this problem.
Note that the values for λ are not needed to solve the original optimization problem even though they may be computed while finding x, y and z.

Exercises

1. Obtain each of the following indicated derivatives.

a) $\dfrac{d}{dx} \displaystyle\int_{1}^{x^2} \cos t^2 \, dt$

b) $\dfrac{d}{dx} \displaystyle\int_{x^3}^{4} \ln(3 + \sqrt{t}) \, dt$

c) $\dfrac{d}{dx} \displaystyle\int_{0}^{\pi/2x} x \sin xt \, dt$

d) $\dfrac{d}{d\alpha} \displaystyle\int_{2\alpha}^{\alpha^3} e^{-\alpha t^2} \, dt$

2. If $\phi(\alpha) = \displaystyle\int_{\sqrt{\alpha}}^{1/\alpha} \sin \alpha x^2 \, dx$, find $\dfrac{d\phi}{d\alpha}$.

3. Verify the result of Example 3. by first integrating with respect to t and
    then differentiating with respect to x.

4. Leibnitz's rule can sometimes be used to evaluate integrals that cannot be
    integrated in a straightforward way. For instance, to evaluate

$F(x) = \displaystyle\int_{0}^{\infty} \dfrac{e^{-t} \sin xt}{t} \, dt$, first differentiate with respect to x to obtain

$\dfrac{dF}{dx}$. The resulting integral can be evaluated to give an expression for

$\dfrac{dF}{dx}$, which may then be integrated to give F(x).

5. Find the critical points for each of the following functions and test for maxima and minima.

a) $f = \sqrt{4 - 2x^2 - y^2}$ b) $f = x^2 - 3xy - y^2$

c) $f = x^3 + x^2y + y^2$ d) $f = x^2 + y^2 + 2y - 9$

e) $f = xy + yz + z^2$ f) $f = 3 - x^4 - 2x^2 - y^2 - 4(z - 1)^2$

6. Find the relative maximum and relative minimum values for
    $f(x,y) = x^4 + y^4 - 2x^2 + 8y^2 + 4$. Does this function have an absolute maximum or an absolute minimum?

7. Find the dimensions of a rectangular box (with top) that has a maximum volume if the surface area is fixed at 2 square units.

8. Find the area of the largest rectangle that can be fit inside the ellipse
    $x^2 + 2y^2 = 4$.

9. Find the critical points of each of the following functions with constraints and determine whether they are optimum points or not.

a) $f = x^2 - y^2$ subject to $x^2 + y^2 = 4$

b) $f = y + z$ subject to $x^2 + y^2 + z^2 = 1$

c) $f = xyz$ subject to $x^2 + z^2 = 1$ and $x - y = 0$
    Hint: Consider $F(x,y,z,\lambda_1,\lambda_2) = xyz + \lambda_1(x^2 + z^2 - 1) + \lambda_2(x - y)$

10. Find the maximum and minimum of $f(x,y,z) = x^3y^2z$ subject to the conditions
    $x + y + z = 6$, $x > 0$, $y > 0$, $z > 0$.

11. Find the shortest distance from the origin to the curve of intersection of the surfaces $xyz = 1$ and $y = x$.

4

MULTIPLE INTEGRATION

4.1 DOUBLE INTEGRALS

Multiple integrals arise in a large variety of applications. The concept of the double integral will be developed here using the example of calculating the volume under a surface $z = f(x,y)$ and above a region R in the xy plane as shown in Figure 1. Basically, we divide the region R into small rectangles with a mesh of straight lines parallel to the coordinate axis as shown in Figure 2. Then the volume under the surface $z = f(x,y)$ above each rectangle is approximated as the area of the rectangle multiplied by a typical height $f(x_i,y_j)$ where x_i,y_j is a point

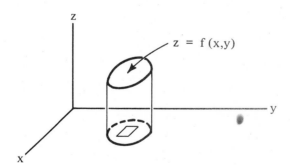

Figure 1.

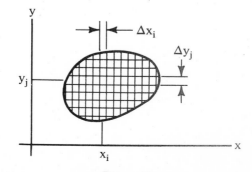

Figure 2.

in the rectangle. Thus the desired total volume will be approximated by

$$V = \sum_i \sum_j f(x_i, y_j) \, \Delta x_i \, \Delta y_j, \tag{1}$$

where the indicated sums are over indices such that the rectangle containing x_i, y_j is not wholly outside R. Equation (1) represents an approximation since for most points in each rectangle $f(x_i, y_j)$ represents only an approximate value for the height and the rectangle containing x_i, y_j may be partially outside R. However as Δx_i gets smaller _and_ as Δy_j gets smaller these approximations become more exact and we thus define the double integral over a region R as:

$$\iint\limits_R f(x,y) \, dA = \lim_{\substack{\Delta x_i \to 0 \\ \Delta y_j \to 0}} \sum_i \sum_j f(x_i, y_j) \, \Delta x_i \, \Delta y_j. \tag{2}$$

The student should review the corresponding definition for areas and single integrals. Equation (2) is sometimes used to numerically evaluate double integrals, however, we are basically interested in working with the left hand side of Eq. (2).

Two properties of integrals will be stated here without proof. The student will recognize the corresponding results for single integrals. The first is the linearity property:

$$\iint\limits_R [af(x,y) + bg(x,y)] dA = a \iint\limits_R f(x,y) dA + b \iint\limits_R g(x,y) dA \tag{3}$$

where a and b are constants and the second is

$$\iint\limits_{R_1} f(x,y) dA + \iint\limits_{R_2} f(x,y) dA = \iint\limits_R f(x,y) dA. \tag{4}$$

In Eq. (4) the region R is the "sum" of the regions R_1 and R_2 whenever R_1 and R_2 have no points in common except perhaps a curve of intersection. Both results can also be extended to triple integrals and thus are used throughout this chapter.

Double integrals as shown on the left of Eq. (2) are evaluated as _iterated integrals_ in practice. If the region R is such that any line parallel to the y axis meets the boundary R in at most two places then

$$\iint\limits_{R} f(x,y)\ dA\ =\ \int_{a}^{b}\int_{g_1(x)}^{g_2(x)} f(x,y)\ dydx,\tag{5}$$

where a, b, $g_1(x)$ and $g_2(x)$ are defined in Figure 3.

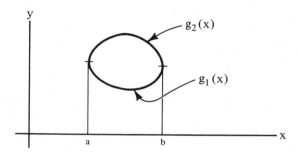

Figure 3.

Example 1.   Evaluate $\iint\limits_{R} xy\ dA$ over the region bounded by

$y = x^2$, $x = 1$ and $y = 0$.

In order to visualize the limits we sketch the given region as shown in Figure 4.   Thus

$$\iint\limits_{R} xy\ dA\ =\ \int_{0}^{1}\int_{0}^{x^2} xy\ dydx,$$

since if x is held constant between 0 and 1, then y will vary

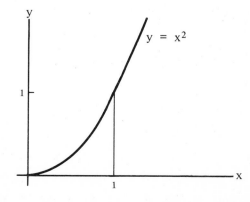

Figure 4.

between 0 and the curve $y = x^2$. Carrying out the indicated integration we get:

$$\iint\limits_{R} xy \ dA \ = \ \int_{0}^{1} x\left(\frac{y^2}{2}\right) \ \Big|_{y=0}^{y=x^2} \ dx$$

$$= \ \int_{0}^{1} \frac{x^5}{2} \ dx \ = \ \frac{1}{12} \ .$$

Notice that the numerical answer in Example 1. represents the volume under $z = xy$ and above the given region. In Eq. (5) and Example 1. we integrate y first, holding x constant, and then integrate with respect to x. The reverse of this is also possible, as shown in

$$\iint\limits_{R} f(x,y) \ dA \ = \ \int_{c}^{d} \int_{h_1(y)}^{h_2(y)} f(x,y) \ dxdy \tag{6}$$

where c, d, $h_1(y)$ and $h_2(y)$ are defined as in Figure 5. Note that the outside limits of integration in both Eqs. (5) and (6) are constants while the inside limits can be functions of either x or y, depending on which variable is the last variable of integration.

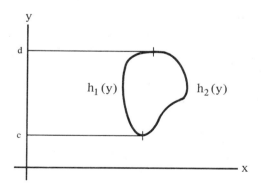

Figure 5.

Example 2.   Rework Example 1. using Eq. (6).

Referring to Figure 4, we see that if y is held constant between 0 and 1, then x varies between $x = \sqrt{y}$ and $x = 1$. Thus

$$\iint\limits_{R} f(x,y) \; dA \;\; = \;\; \int_{0}^{1} \int_{\sqrt{y}}^{1} xy \; dxdy$$

$$= \;\; \int_{0}^{1} \frac{x^{2}}{2} \; \Big|_{x=\sqrt{y}}^{x=1} \; y \; dy$$

$$= \;\; \int_{0}^{1} (\frac{1}{2} - \frac{y}{2}) \; y \; dy = \frac{1}{12} \; .$$

In Eq. (6) and Example 2. we have integrated x first and then y. This can always be done if lines parallel to the x axis cut the boundary R in at most two places. In comparing Examples 1. and 2, we see that the same result is obtained. This is always true -- that is, the double integral as defined in Eq. (2) will have the same value no matter which variable is integrated first -- provided the double integral exists.

If the integrand is taken as the constant function, $f(x,y) = 1$, over the whole region R, then $\iint\limits_{R} dA$ represents the area of the region R.

Example 3.   Find the area of the region bounded by $y = x$ and $y = x^{2}$.

Since the curves $y = x$ and $y = x^{2}$ intersect at the points $(0,0)$ and $(1,1)$, we have

$$A \;\; = \;\; \int_{0}^{1} \int_{x^{2}}^{x} dydx \tag{7}$$

$$= \;\; \int_{0}^{1} (x - x^{2})dx$$

$$= \;\; \frac{1}{2} - \frac{1}{3} = \frac{1}{6} \; .$$

Again, the limits in Eq. (7) can be verified by sketching the region.

In many applications of double integrals, the integral to be evaluated is already given as an iterated integral. If the evaluation is straightforward there is no problem. However, if the integration is complicated or not possible for some reason, an alternative is offered by the idea of <u>changing</u> <u>the</u> <u>order</u> <u>of</u> <u>inte</u>-

gration. This concept is shown in the following examples.

Example 4.   Interchange the order of integration for the integral

$$\int_0^1 \int_{y^2}^1 x^2 \, dxdy.$$

In order to interchange the order of integration we first sketch the
region R as defined by the limits on the given integral, as shown
in Figure 6. Thus we have

$$\int_0^1 \int_{y^2}^1 x^2 \, dxdy \; = \; \int_0^1 \int_0^{\sqrt{x}} x^2 \, dydx.$$

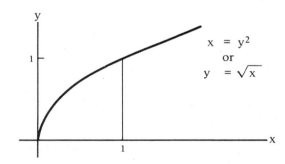

$$x = y^2$$
$$\text{or}$$
$$y = \sqrt{x}$$

Figure 6.

Example 5.   The integral $\displaystyle\int_0^x \int_0^t f(u) \, dudt$ may be partially

evaluated since the integrand is independent of t. To do this, the
order of integration is interchanged with the help of Figure 7. to

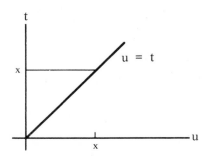

$$u = t$$

Figure 7.

give:

$$\int_0^x \int_0^t f(u) \, du \, dt = \int_0^x \int_u^x f(u) \, dt \, du$$

$$= \int_0^x (x - u) f(u) \, du.$$

As shown in Examples 4. and 5, the idea behind the interchange of the order of integration is to sketch the region R in the "xy" plane and then set up the limits in the reverse order. In some examples, the new limits may necessitate writing the double integral as the sum of two (or more) integrals as shown in the next example.

Example 6.   Interchange the order of integration for

$$\int_{-1}^2 \int_{x^2}^{x+2} f(x,y) \, dy \, dx$$

Again, a sketch of the region is imperative and is shown in Figure 8.

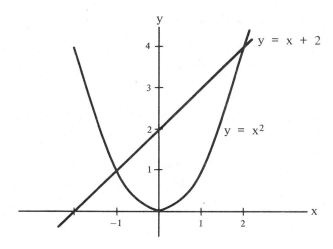

Figure 8.

If the order of integration is reversed, then y is held constant while x is varied. The lower limit on x then depends on what y value is chosen since in this case there are two lower curves, $x = -\sqrt{y}$ for $0 \le y \le 1$ and $x = y - 2$ for $1 \le y \le 4$. Thus the sum

$$\int_0^1 \int_{-\sqrt{y}}^{+\sqrt{y}} f(x,y) \, dxdy + \int_1^4 \int_{y-2}^{\sqrt{y}} f(x,y) \, dxdy$$

represents the same value as the original double integral.

Notice that in all three of the last examples that the integrand has not changed when the order of integration has been reversed.

In some applications the idea of squaring an integral (or more generally multiplying two integrals) arises. The procedure for doing this is illustrated in the next example.

Example 7.
$$\left(\int_0^1 f(x) \, dx \right)^2 = \left(\int_0^1 f(x) \, dx \right) \left(\int_0^1 f(x) \, dx \right)$$

$$= \left(\int_0^1 f(x) \, dx \right) \left(\int_0^1 f(y) \, dy \right)$$

$$= \int_0^1 \int_0^1 f(x)f(y) \, dxdy.$$

What we have seen in Example 7. is that the variable of integration is really any parameter, thus the multiplication process can be exchanged for the evaluation of an iterated or double integral. Notice that the limits in Example 7. are not functions, but constants, which is always true when integrals are multiplied.

Iterated integrals also arise when the concept of integration under the integral sign is considered. For instance if $\phi(t)$ in the integral $\int_a^b \phi(t)dt$ is given as:

$$\phi(t) = \int_{u_1(t)}^{u_2(t)} f(x,t) \, dx \tag{8}$$

then

$$\int_a^b \phi(t) \, dt = \int_a^b \int_{u_1(t)}^{u_2(t)} f(x,t) \, dxdt. \tag{9}$$

Here again, the difference in the integration variables is important. Equation (9) we see is an iterated integral and hence can be treated as such.

In conclusion, we have introduced the concept of a double integral through the idea of calculating volumes. These double integrals are evaluated through the use of iterated integrals and several concepts in the use and manipulation of these iterated integrals have been presented. There are, of course, a wide variety of applications of double integrals, but they will not be discussed here as the variety of details are beyond the scope of this text.

Exercises

1. Evaluate each of the following double integrals and sketch the region over which the integration extends.

a) $\displaystyle\int_{1}^{2}\int_{y}^{y^2} dxdy$

b) $\displaystyle\int_{0}^{\pi}\int_{0}^{\sin x} y\, dydx$

c) $\displaystyle\int_{0}^{\pi}\int_{0}^{x} x\sin(y)\, dydx$

d) $\displaystyle\int_{-1}^{1}\int_{y}^{1} (x-y)^2\, dxdy$

2. Write an equivalent double integral with the order of integration reversed for each of the following integrals.

a) $\displaystyle\int_{0}^{\sqrt{2}}\int_{-\sqrt{4-2y^2}}^{\sqrt{4-2y^2}} y\, dxdy$

b) $\displaystyle\int_{0}^{3}\int_{1}^{e^x} xy\, dydx$

c) $\displaystyle\int_{-2}^{1}\int_{x^2+4x}^{3x+2} dydx$

d) $\displaystyle\int_{-2}^{1}\int_{y-1}^{1-y^2} e^{-x^2}\, dxdy$

3. Evaluate each of the following integrals.

a) $\displaystyle\iint_{R} (x^2+y^2)dA$ when R is the region inside the triangle with vertices $(0,0),(1,1),(0,1)$

b) $\displaystyle\iint_{R} xy^2\, dA$ when R is the region inside the right half of the unit circle $x^2+y^2=1$

   (Hint: consider carefully which order of integration is appropriate)

4. Find the area of the region bounded by the semi-circle $y = \sqrt{a^2 - x^2}$ and the lines $x = \pm a$, $y = a$, using double integration.

5. Find the volume of the tetrahedron bounded by $2x + y + z = 2$ and the coordinate planes using double integration.

6. Find the volume in the first octant bounded by the cylinders $x^2 + y^2 = a^2$ and $x^2 + z^2 = a^2$ using double integrals.

7. The following exercises are simple illustrations of the importance of a full understanding of the nature of the integration variable.

   a) Write $\left(\int_0^1 x \, dx \right)^2$ as a double or iterated integral and evaluate.

   b) If $F(x) = \int_0^2 \sin(t-x)dt$, find $F(t)$.

   c) If $\phi(\alpha) = \int_0^\alpha \sin(x)dx$, find $\int_0^\pi \phi(x)dx$, using iterated integrals.

8. Starting with the result $\int_0^{2\pi} (\alpha - \sin x)dx = 2\pi\alpha$, use Eqs. (8) and (9) to

   show that

   $$\int_0^{2\pi} [(b - \sin x)^2 - (a - \sin x)^2]dx = 2\pi(b^2 - a^2)$$

   for all constants a and b.

4.2 TRIPLE INTEGRALS AND CHANGE OF VARIABLES

In the last section the concept of a double integral was discussed in detail. All that work can easily be extended to triple (or more) integrals, as we shall show first here.

The concept of a triple integral can be obtained from thinking of a three-dimensional object whose density varies from point to point. If the object is thought of as being composed of many box shaped volumes of dimensions Δx_i, Δy_j, and Δz_k, then the approximate mass of each little box will be given by

$$\Delta M = d(x_i, y_j, z_k) \Delta x_i \Delta y_j \Delta z_k, \tag{1}$$

where $d(x,y,z)$ is the density. Again, this is an approximation since d is evaluated at only one point (x_i, y_j, z_k) as a typical density from the whole box. However, as Δx_i, Δy_j, and Δz_k all go to zero, then $d(x_i, y_j, z_k)$ will be a better and better approximation. Thus the total mass of the volume (after summing) will be given by

$$M = \lim \sum_i \sum_j \sum_k d(x_i, y_j, z_k) \Delta x_i \Delta y_j \Delta z_k, \tag{2}$$

where as before the sums are over indices such that the box containing (x_i, y_j, z_k) is not wholly outside the given three dimensional object and the limit is taken as Δx_i, Δy_j, and Δz_k all go to 0. The term on the right of Eq. (2) is defined as the triple integral of $d(x,y,z)$ over the given volume:

$$M = \iiint_V d(x,y,z) \ dV \tag{3}$$

In arriving at Eqs. (2) and (3) we used the concept of mass and density. However, there are many other possible interpretations. If $d(x,y,z) = 1$ for all points in V, then the integral is simply the volume of V. First moments and moments of inertia may also be found by using an appropriate integrand.

The iterated integrals associated with the triple integral defined in Eq. (3) are obtained by the following procedure:

(1) Suppose V is bounded below by a surface $z = f_1(x,y)$ and above by the surface $z = f_2(x,y)$.

(2) Let R denote the region in the xy plane which is the orthogonal projection of the solid V onto the xy plane.

Then the equation

$$\iiint_V d(x,y,z) \ dV = \iint_R \int_{f_1(x,y)}^{f_2(x,y)} d(x,y,z) dz \ dA, \tag{4}$$

reduces a triple integral to an iterated integral. Step (1) assumes any line parallel to the z axis cuts the surface bounding the volume in at most two places.

Example 1.   Find the integral representing the volume enclosed by
                      the two surfaces $z = 8 - x^2 - y^2$ and $z = x^2 + 3y^2$.

    A rough sketch of the surfaces in the first octant is shown in Figure 1.

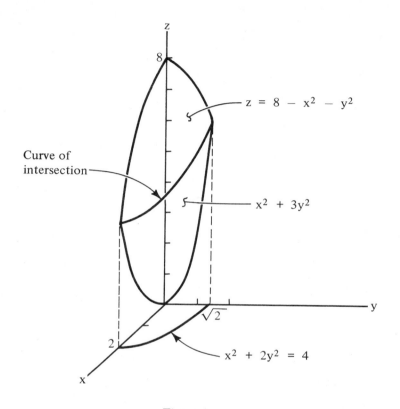

Figure 1.

From the sketch in Figure 1, we see that the lower surface is $z = x^2 + 3y^2$
while the upper surface is $z = 8 - x^2 - y^2$. To find the limits of integration for the xy variables we project the volume onto the xy plane by
eliminating z from the two given equations. This yields the equation
$x^2 + 2y^2 = 4$, which can be interpreted as the curve bounding the region
R. Thus

$$V \;=\; 4 \int_{0}^{\sqrt{2}} \int_{0}^{\sqrt{4-2y^2}} \int_{x^2+3y^2}^{8-x^2-y^2} dz\,dx\,dy, \tag{5}$$

where symmetry has been used so that the limits in the xy plane only
cover the region in the first quadrant.

Equation (4) shows one way of evaluating a triple integral as an iterated in-

tegral. There are two others obtained by projecting the volume on the xz or the yz
planes. Of course, each of these has two possible ways of continuing, as for example in Eq. (5) we could have integrated y before x, as we saw in the last section.
Which order of integration is used depends very much on the surfaces and how they
project onto the coordinate planes. As with double integrals, the limits of integration follow a pattern which is quite important. By referring to Eqs. (4) and (5)
we see that the limits on the last integration must be constants, while the limits
on the middle integration can be functions of the last variable of integration, and
the limits on the first integration can be functions of both the remaining integration
variables.

In the evaluation of double and triple integrals in practical applications it
is often imperative to change the variable of integration. The basic relationship
in this case is given by:

$$\iiint\limits_{V} f(x,y,z)\ dV \ = \ \iiint\limits_{V'} F(u,v,w) \left| \frac{\partial(x,y,z)}{\partial(u,v,w)} \right|\ dudvdw, \qquad (6)$$

where $x = x(u,v,w)$, $y = y(u,v,w)$, $z = z(u,v,w)$ is the transformation,
$f[x(u,v,w),y(u,v,w),z(u,v,w)] = F(u,v,w)$, V' is the transformed region, and

$\frac{\partial(x,y,z)}{\partial(u,v,w)}$ is the Jacobian as discussed earlier in Section 3.2. We will not derive
Eq. (6), but will illustrate it for the common transformed coordinates: polar,
cylindrical and spherical.

For polar and cylindrical coordinates we have the relations

$$
\begin{aligned}
x &= r\ \cos\ \theta \\
y &= r\ \sin\ \theta \\
z &= z
\end{aligned}
\qquad (7)
$$

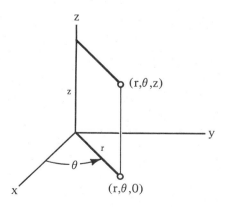

Figure 2.

as can be verified from Figure 2. Thus

$$\frac{\partial(x,y,z)}{\partial(r,\theta,z)} = \begin{vmatrix} \cos\theta & -r\sin\theta & 0 \\ \sin\theta & r\cos\theta & 0 \\ 0 & 0 & 1 \end{vmatrix} = r. \tag{8}$$

Example 2.   Change the double integral $\displaystyle\int_0^{a/\sqrt{2}} \int_y^{\sqrt{a^2-y^2}} x^2\, dxdy$

to an equivalent double integral in terms of polar coordinates.

In order to find the correct limits it is imperative to sketch the region over which the integration is carried out. For this example the region is shown in Figure 3. Thus we have

$$\int_0^{a/\sqrt{2}} \int_y^{\sqrt{a^2-y^2}} x^2\, dxdy = \int_0^{\pi/4} \int_0^{a} (r^2\cos^2\theta)rdrd\theta,$$

since the region is bounded by $\theta = 0$ and $\theta = \pi/4$ as r goes from 0 to a.

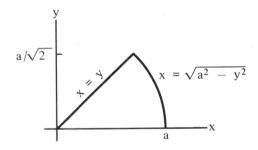

Figure 3.

It should be pointed out that when integration variables are changed the integrand does in general change also. This is not the case when the order of integration is reversed. The extension of Example 2. to a triple integral with cylindrical coordinates would be very similar since z in rectangular and cylindrical coordinates is the same and thus really only x and y are transformed.

For spherical coordinates we have the relations

$$x = \rho\sin\phi\cos\theta$$
$$y = \rho\sin\phi\sin\theta \tag{10}$$
$$z = \rho\cos\phi$$

as can be verified from Figure 4. Thus

$$\frac{\partial(x,y,z)}{\partial(\rho,\phi,\theta)} = \begin{vmatrix} \sin\phi\cos\theta & \rho\cos\phi\cos\theta & -\rho\sin\phi\sin\theta \\ \sin\phi\sin\theta & \rho\cos\phi\sin\theta & \rho\sin\phi\cos\theta \\ \cos\phi & -\rho\sin\phi & 0 \end{vmatrix} \tag{11}$$

$$= \rho^2\sin\phi \;.$$

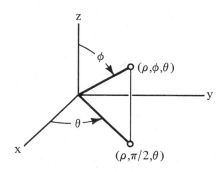

Figure 4.

Example 3.   Find the integral representing the volume of the region bounded above by the sphere $x^2 + y^2 + z^2 = a^2$ and below by the cone $z = \sqrt{x^2+y^2}$.

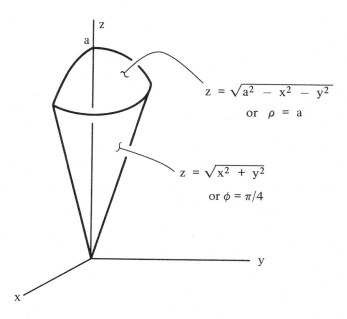

Figure 5.

This example is most easily worked using spherical coordinates since the given sphere is ρ = a and the given cone is ϕ = $\pi/4$ in spherical coordinates, as shown in Figure 5.   Thus

$$V = \int_0^{2\pi} \int_0^{\pi/4} \int_0^a \rho^2 \sin\phi d\rho d\phi d\theta, \tag{12}$$

since dV = $\rho^2 \sin\phi d\rho d\phi d\theta$ from Eqs. (6) and (11).

Notice that the limits in Eq. (12) are obtained first by holding ϕ and θ constant (for the ρ limits) and then holding θ constant (for the ϕ limits).   In general, the limits can be obtained from the sketch in the xyz coordinates rather than drawing the corresponding region in the r,θ,z or ρ,ϕ,θ coordinates.   The student should set up the volume integral for Example 3. using rectangular coordinates and verify that indeed spherical coordinates are much simpler.

Examples 2. and 3. illustrate the use of Eq. (6) for polar, cylindrical and spherical coordinates.   In practice Eq. (6) may be used for any change of coordinates (provided the Jacobian is not zero) as shown in the following example.

Example 4.   Evaluate the integral $\iint_R$ x dA when   R   is the

parallelogram shown in Figure 6.

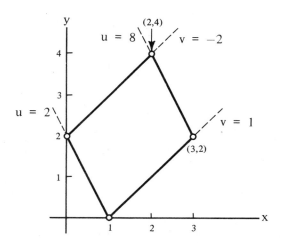

Figure 6.

From the indicated points, we see that the sides of R are straight lines of the form

$$2x + y = c_1 \text{ and } x - y = c_2$$

for appropriate choices of c_1 and c_2. We therefore introduce the new coordinates

$$u = 2x + y \text{ and } v = x - y, \tag{13}$$

so that R then corresponds to the rectangle $2 \le u \le 8$, $-2 \le v \le 1$ in the u,v plane. Solving Eqs. (13) for x and y in terms of u and v we obtain

$$x = \frac{1}{3}(u + v) \text{ and } y = \frac{1}{3}(u - 2v). \tag{14}$$

Hence Eq. (6) yields

$$\iint_R x \, dA = \int_{-2}^{1} \int_{2}^{8} x(u,v) \left| \frac{\partial(x,y)}{\partial(u,v)} \right| du \, dv$$

$$= \int_{-2}^{1} \int_{2}^{8} \frac{1}{3}(u + v) \left| -\frac{1}{3} \right| du \, dv,$$

where Eqs. (14) have been used for x and for the calculation of the Jacobian. Thus

$$\iint_R x \, dA = \frac{1}{9} \int_{-2}^{1} \int_{2}^{8} (u + v) \, du \, dv$$

$$= \frac{1}{9} \int_{-2}^{1} \left(\frac{u^2}{2} + uv \right) \Bigg|_{u=2}^{u=8} dv$$

$$= \frac{1}{9} \int_{-2}^{1} (30 - 6v) dv = 11.$$

In Example 4, the change of variables was chosen to simplify the limits of integration. In other cases it may be more appropriate to choose a change of variables to facilitate the integration of the integrand.

Exercises

1. Evaluate each of the following integrals,

a) $\displaystyle\int_0^2 \int_1^3 \int_1^2 xy^2z \; dzdydx$ b) $\displaystyle\int_0^1 \int_0^x \int_{xy}^2 xyz \; dzdydx$

2. Find by triple integration:

   a) The volume bounded by the plane $x + 2y + 3z = 6$ and the coordinate planes.
   b) The volume bounded by the cylinder $x^2 + y^2 = 1$ and the planes $z = 0$ and $z = x$.

3. Consider a triple integral representing the volume in the first octant bounded below by the plane $z = 0$, laterally by the "cylinder" $x^2 + 4y^2 = 4$ and above by the plane $z = x + 2$.

   a) Write an iterated integral integrating with respect to z first.
   b) Write an iterated integral integrating y first.
   c) Write an iterated integral integrating x first.
   Do not evaluate these.

4. Evaluate the integral $I = \displaystyle\int_0^\infty e^{-x^2} \; dx$ by the following steps:

   a) Write I^2 as an iterated integral, using y as the second variable of integration.
   b) Change the variables of integration to polar coordinates.
   c) Evaluate the double integral in polar coordinates.
   d) Find I from part c.

5. Use cylindrical coordinates to find the volume cut from the sphere $x^2 + y^2 + z^2 = 4$ by the cylinder $x^2 + y^2 = 1$.

6. Use cylindrical coordinates to evaluate the integral $\iiint\limits_{V} \sqrt{x^2 + y^2 + z^2}\, dV$

    when V is the volume bounded by the plane $z = 3$ and the cone $z = \sqrt{x^2 + y^2}$.

7. Use spherical coordinates to evaluate the integral $\iiint\limits_{V} (x^2 + y^2)\, dV$ when

    V is the volume bounded by the concentric spheres $x^2 + y^2 + z^2 = 1$ and $x^2 + y^2 + z^2 = 4$.

8. If the density of a spherical ball of radius a is proportional to the distance from the center, find its total mass using spherical coordinates. (Hint: density = $k\rho$, where k is a constant).

9. Evaluate the integral $\iint\limits_{R} (x - y)^2\, dA$ when R is the region bounded by

    the four lines $x + 2y = 2$, $x + 2y = 4$, $x - y = 1$ and $x - y = -2$.

10. Evaluate the integral $\displaystyle\int_{0}^{1} \int_{0}^{1-x} \sin\left(\frac{y}{x+y}\right) dy\, dx$ by using the change of variables

    $u = \frac{y}{x+y}$ and $v = x + y$.

5

VECTOR ANALYSIS

5.1 THE GEOMETRY OF VECTORS

There are many physical concepts arising in engineering that require more than a number to describe them. The velocity of an automobile is given as 30 miles per hour in the westerly direction, a force of 15 lbs. in the vertical direction, etc. Such sets of numbers are called <u>vectors</u>. A vector quantity is composed of two parts: a <u>scalar</u> or number, called the <u>magnitude</u>, and a <u>direction</u> in space. Thus the vector concept is closely related to the geometrical idea of a <u>directed line segment</u>, which is used to represent a vector in many diagrams.

The above geometrical definition of a vector will be used at the start to define equality, addition and subtraction of vectors since in many physical applications these concepts are very useful. After these are given, a more rigorous rectangular components form of a vector will be presented that is more easily handled mathematically.

We will denote vectors by single bold faced letters: A, v and e_1 and their magnitudes will be denoted as $|A|$, $|v|$ and $|e_1|$. Two vectors are said to be <u>equal</u>, $A = B$, if they have the same direction <u>and</u> same magnitude.

Example 1.   In Figure 1, A and B have the same direction and magnitude and hence they are equal, while C cannot be equal to A or B since it does not have the same direction.

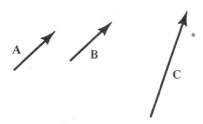

Figure 1.

The <u>zero</u> vector, 0, has zero magnitude and no <u>defined</u> <u>direction</u>.

The geometrical definition of the sum of two vectors follows from the composition of two forces acting on a body and is shown geometrically by the solid lines in Figure 2. This definition is also known as the <u>parallelogram</u> addition of two vectors. Since the sides of a parallelogram have the same magnitude and direction it is clear that the commutative law holds:

$$A + B = B + A , \tag{1}$$

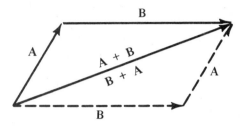

Figure 2.

which is illustrated in Figure 2, by dotted lines. The associative law

$$(A + B) + C = A + (B + C) \tag{2}$$

also holds as shown in Figure 3. Equation (2) says there is no ambiquity in writing

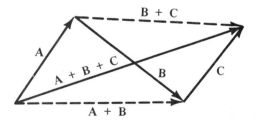

Figure 3.

A + B + C without parenthesis.

A very important inequality in many applications is

$$|A + B| \leq |A| + |B| , \tag{3}$$

which is known as the <u>triangle</u> <u>inequality</u>. It can be justified very easily using the parallelogram addition of Figure 2, since the length of one side of a triangle is less than or equal to the sum of the lengths of the other two. In Eq. (3), equality holds only if A and B have the same direction.

We define -B to be a vector with the same magnitude but opposite direction to that of B. We thus can define vector subtraction as:

$$A - B = A + (-B), \tag{4}$$

which is shown geometrically in Figure 4.

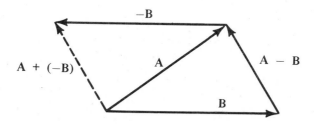

Figure 4.

If s is any real number, then sA, the scalar multiple of A, is a vector in the same direction as A if s is positive and opposite direction if s is negative. The magnitude of sA is $|s||A|$, where $|s|$ is the absolute value of the real number s.

Example 2.   In Figure 5, the vector B is (2/3)A since it has the same direction as A but is only two thirds as long. Likewise C is (-1/2)A since it has the opposite direction and is half as long.

Figure 5.

A unit vector is a special vector that has a magnitude of one unit in a specified direction. Three particularly useful unit vectors are the vectors i, j and k with directions respectively along the positive x, y, z axis in a rectangular coordinate system as shown in Figure 6. For an arbitrary nonzero vector A, the vector A/|A| denotes the unit vector in the same direction as A.

Now let A be any vector in three dimensional space. If we position A at the origin of Figure 6. with its direction away from the origin then we may project the line segment representing A onto each of the coordinate axis to obtain its <u>components</u>

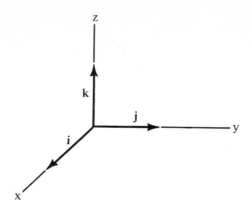

Figure 6.

a_1, a_2 and a_3 along the coordinate axis. Referring to Figure 7. and using the

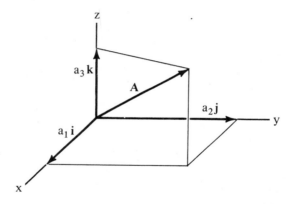

Figure 7.

parallelogram law of addition we see that

$$A = a_1 i + a_2 j + a_3 k,\qquad\qquad(5)$$

which is called the <u>component</u> representation of the vector A. The length or magnitude
of A can be found by the Pythagorean theorem and is given by

$$|A| = \sqrt{a_1^2 + a_2^2 + a_3^2}\ .\qquad\qquad(6)$$

The direction of A is specified by the angles α, β and γ that the line segment in
Figure 7. makes with the positive x, y, and z axis respectively. Geometrically the
situation may be seen in Figure 8. and analytically the angles are given by:

$$\cos\alpha = \frac{a_1}{|A|},\ \cos\beta = \frac{a_2}{|A|}\ \text{and}\ \cos\gamma = \frac{a_3}{|A|}.\qquad(7)$$

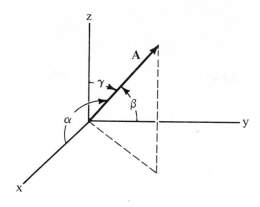

Figure 8.

Equations (7) can be found by considering the plane containing the vector A and the relevant positive axis. For example, to find cos α consider the plane formed by A and i, as shown in Figure 9. Elementary trigonometric relations give the first

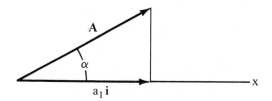

Figure 9.

of Eqs. (7) and similarly for the others.

Example 3.   Find the magnitude and the direction cosines for

$$A = i - j + 3k.$$

The magnitude is given by Eq. (6) as:

$$|A| = \sqrt{1 + 1 + 9} = \sqrt{11},$$

and the direction cosines are given by Eqs. (7) as:

$$\cos \alpha = \frac{1}{\sqrt{11}}, \quad \cos \beta = \frac{-1}{\sqrt{11}}, \quad \text{and} \quad \cos \gamma = \frac{3}{\sqrt{11}}.$$

From Eqs. (7) and (6) we obtain the following:

$$\cos^2\alpha + \cos^2\beta + \cos^2\gamma = 1 \qquad (8)$$

which says that once two angles are given, the third is no longer arbitrary. However, the third angle is not uniquely described as shown in the following example.

Example 4.   If $\alpha = 60°$ and $\gamma = 60°$, find β .

We know that $\cos 60° = \frac{1}{2}$ so that

$$\cos^2\beta = 1 - \cos^2\alpha - \cos^2\gamma$$

$$= 1 - \frac{1}{4} - \frac{1}{4} = \frac{1}{2} .$$

Thus $\cos\beta = \pm\sqrt{2}/2$ and hence

$$\beta = 45° \text{ or } 135° .$$

Exercises

1. Given the two dimensional vectors, $A = i + 2j$ and $B = i - j$, find each of the following vectors using a geometrical drawing.

a) $A + B$ b) $A - B$

c) $2A + 4B$ d) $B - 2A$

e) $.5A + B$ f) $.5B - 1.5A$

2. Using a geometrical proof, show that if $A + B = C$, then $B = C - A$.

3. A man travels southeast 20 miles, due west 5 miles and due south 8 miles. Use an appropriate scale to graphically determine how far and in what direction he is from his starting position.

4. If A, B, C and D are vectors from the origin to the points A, B, C, D in the plane and if $B - A = C - D$, then is ABCD a parallelogram? Does the result hold if A, B, C and D are points in 3 space?

5. Use the triangle inequality to show that $|A - B| \geq |A| - |B|$.

6. Let A = 2i – 4j + k.

    a) Find $|A|$.

    b) Find cos α, cos β, cos γ.

    c) Verify Eq. (8).

    d) Find a unit vector in the direction of A.

7. Repeat Exercise 6. for A = 5j + 6k.

8. a) Determine γ if cos α = $1/\sqrt{2}$ and cos β = $1/\sqrt{2}$.

    b) Determine α if cos β = 1/2 and cos γ = $1/\sqrt{2}$.

9. The following short answer questions provide a summary of the geometrical concepts introduced in this section.

    a) If $|A|$ = $|B|$ is it necessarily true that A = B?

    b) Can $|A|$ < 0?

    c) If $|A|$ = 1.5, what are $|2A|$ and $|-3A|$?

    d) How many distinct vectors of unit magnitude are perpendicular to a given plane?

    e) How many distinct vectors of unit magnitude are perpendicular to a given line?

    f) If B is a scalar multiple of A, is A a scalar multiple of B?

5.2 COMPONENTS AND THE DOT PRODUCT

In the last section we introduced the concept of a vector through its geometric use. At the end of the section we related the geometric vector to its component form, and found its magnitude and direction cosines in terms of its components. In this section we will extend the use of the component form to other operations with vectors. The component form is much easier to use, but for many applications the geometric concepts will certainly help in formulating results and ideas.

Vector addition and multiplication by a scalar in component form are given by:

$$A + B = (a_1 + b_1)i + (a_2 + b_2)j + (a_3 + b_3)k \qquad (1)$$

and

$$sA = sa_1 i + sa_2 j + sa_3 k. \qquad (2)$$

The illustrations in Figures 1. and 2. justify Eqs. (1) and (2) using the geometric definitions of Section 5.1. Only two dimensions are illustrated, but the extension to three dimensions is straightforward.

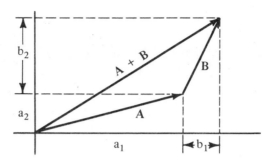

Figure 1.

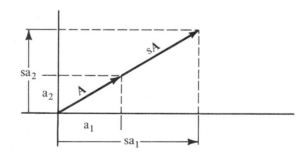

Figure 2.

Example 1.   If $A = 2i + 3j - k$ and $B = i - j + 2k$,
then

$$2A = 4i + 6j - 2k$$

and

$$2A + B = 5i + 5j.$$

The component form of a vector allows us to use vectors to solve many geometrical problems. The vector from the origin to the point (x_0, y_0, z_0) in three dimensional space is given by:

$$R_0 = x_0 i + y_0 j + z_0 k \qquad (3)$$

since obviously x_0, y_0 and z_0 are the components of R_0 in the i, j and k directions. Such vectors are frequently called <u>position vectors</u>. The vector going from the point (x_0, y_0, z_0) to the point (x_1, y_1, z_1) is given by:

$$R_1 - R_0 = (x_1 - x_0)i + (y_1 - y_0)j + (z_1 - z_0)k \tag{4}$$

again since the difference in coordinates yields the desired components as seen in Figure 3. The equation of a straight line through two points is now easily obtained

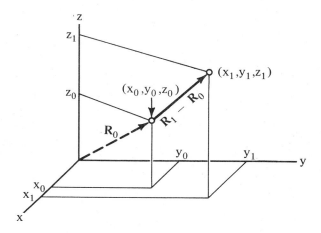

Figure 3.

using Eqs. (3) and (4). From Figure 4. it can be seen that any point (x,y,z) on the

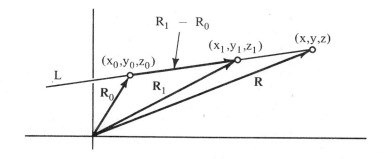

Figure 4.

line L determined by the points (x_0, y_0, z_0) and (x_1, y_1, z_1) is given by

$$R = R_0 + t(R_1 - R_0), \tag{5}$$

where $R = xi + yj + zk$ and R_0 and R_1 are found by using Eq. (3) appropriately. For different values of t we obtain different points on the line. Equation (5) represents the <u>vector</u> equation of the straight line. The parametric equations of a straight line are obtained by equating components in Eq. (5):

$$x = x_0 + t(x_1 - x_0)$$

$$y = y_0 + t(y_1 - y_0) \tag{6}$$

$$z = z_0 + t(z_1 - z_0) \ .$$

Example 2.   Find the vector and the parametric equations of the line
                      through (1,2,-3) and (2,1,1).

In this case, $R_0 = i + 2j - 3k$ and $R_1 = 2i + j + k$, and thus

$$R = (i + 2j - 3k) + t(i - j + 4k)$$

and

$$x = 1 + t$$
$$y = 2 - t$$
$$z = -3 + 4t.$$

The scalar product of two vectors is a real number defined by:

$$A \cdot B = |A||B|\cos \theta, \tag{7}$$

where θ is the angle between the two vectors A and B. This is also called the dot
product since a dot is almost always used as in Eq. (7). In a few simple cases, the
scalar product is found from Eq. (7) as shown in Figure 5. However, for general

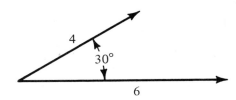

Figure 5.

vectors the component form is more appropriate:

$$A \cdot B = a_1 b_1 + a_2 b_2 + a_3 b_3, \tag{8}$$

which can be obtained from Eq. (7) by finding $\cos \theta$ in terms of the components of
A and B.

Example 3.   If $A = 2i - j + 3k$ and $B = i + 3j$, then

$$A \cdot B = 2 \cdot 1 + (-1)(3) + 3 \cdot 0 = -1.$$

Example 4.   Find the angle between the vectors of Example 3.

Solving Eq. (7) for $\cos \theta$ we obtain

$$\cos \theta = \frac{A \cdot B}{|A| \, |B|} = \frac{-1}{\sqrt{14} \, \sqrt{10}} \quad .$$

Thus

$$\theta = \cos^{-1} \frac{-1}{\sqrt{140}}$$

so that the desired angle is approximately 94°51'.

Example 5.   Find the projection of $A = 2i + j + k$ on $B = i - j + 3k$.

From Figure 6. the projection of A on B is given by $|A| \cos \theta$ and from Eq. (7) we have

$$|A| \cos \theta = \frac{A \cdot B}{|B|}$$

$$= \frac{2 - 1 + 3}{\sqrt{11}} = \frac{4}{\sqrt{11}} \quad .$$

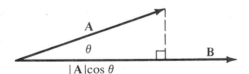

Figure 6.

One important property of the dot product is obtained from Eq. (7) when A and B are perpendicular to each other. In this case $\theta = \pi/2$ and thus $A \cdot B = 0$. The converse is also true provided that A and B are non-zero, vectors. That is, $A \cdot B = 0$ implies that A and B are perpendicular provided neither A nor B is the zero vector. Some other properties of the dot product are:

$$A \cdot B = B \cdot A \tag{9}$$

$$(sA) \cdot (B) = s(A \cdot B) \tag{10}$$

$$A \cdot (B + C) = A \cdot B + A \cdot C \tag{11}$$

$$|A|^2 = A \cdot A , \tag{12}$$

which are easily verified using either Eq. (7) or Eq. (8).

The dot product can be used to find the equation of a plane passing through the point (x_0, y_0, z_0) and perpendicular to a given nonzero vector

$$N = ai + bj + ck. \tag{13}$$

If we let $R = xi + yj + zk$ represent any point on the plane, then $R - R_0$ lies in the plane and must be perpendicular to N, as shown in Figure 7. Thus

$$(R - R_0) \cdot N = 0, \tag{14}$$

on the basis of the above discussion. Using Eq. (8) to rewrite Eq. (14) in terms of scalars we obtain

$$a(x - x_0) + b(y - y_0) + c(z - z_0) = 0, \tag{15}$$

as the scalar equation of the plane. Equation (15) should be compared to the form of a plane that was presented in Section 3.1.

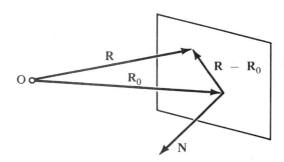

Figure 7.

Example 6.   Find the equation of the plane through $(1,2,-1)$ parallel to the plane $2x + 4y - z = 6$.

Comparing the given plane to Eq. (15) we see that

$$N = 2i + 4j - k$$

is a vector perpendicular to the given plane. Since this vector must also be perpendicular to the desired plane, we get

$$2(x - 1) + 4(y - 2) - (z + 1) = 0$$

as the equation of the desired plane. This last equation may be rewritten as

$$2x + 4y - z = 11,$$

which is a more familiar form.

Exercises

1. Let $A = i + 3j + k$, $B = 2i + 2j - k$ and $C = i - 2j$.

    a) Find $2A - B$.   b) Find $B \cdot C$.

    c) Find $B \cdot (A + C)$.   d) Find $|A + B|$.

    e) Find   s   so that $|sB| = 1$.   f) Find $\cos \theta$ for A and B .

    g) Find the component of B in the direction of A.

2. Find vector and parametric equations of the straight line

    a) passing through the points $(1,3,1)$ and $(-2,1,3)$.

    b) passing through the points $(-2,0,5)$ and $(-2,1,2)$.

    c) passing through the point $(1,2,5)$ and parallel to the vector
        $i + j + 3k$.

    d) passing through the point $(-1,3,2)$ and perpendicular to the plane
        $x - y + 2z = 10$.

    e) passing through $(1,0,1)$ and parallel to the   z   axis.

3. Let A and B be two points with position vectors A and B respectively. Show
    that the line passing through these points may be written in the vector form
    $R = sA + tB$, where $s + t = 1$.

4. A force $F = 5i + 2j - k$ moves a mass from the point $(1,2,-1)$ to the point
    $(2,2,1)$. Find the work done ($w = F \cdot d$, where d is the displacement vector).

5. Find the equation of the planes satisfying each of the following conditions.

    a) The plane passing through $(1,2,2)$ perpendicular to the vector $2\mathbf{i} - \mathbf{j} + 3\mathbf{k}$.

    b) The plane passing through the origin parallel to the plane $x + 2y - 3z = 4$.

    c) The plane passing through $(-1,2,-9)$ perpendicular to the line given by $x = 1 - t$, $y = 2 + 3t$ and $z = -1 + 2t$.

    d) The plane passing through $(1,1,2)$ perpendicular to the y axis.

6. Find a unit vector perpendicular to the plane $2x + y + z = 16$.

7. If $A \cdot B = B \cdot C$ is it necessarily true that $A = C$? Explain your answer.

8. Describe the locus of points R which satisfy the equation $(R - A) \cdot R = 0$ in two and three-dimensional space.

9. Show that $\dfrac{|ax_0 + by_0 + cz_0 - d|}{(a^2 + b^2 + c^2)^{1/2}}$ gives the distance from (x_0, y_0, z_0) to the plane $ax + by + cz = d$.

10. Use the result of Exercise 9. to find the distance from $(1,-1,3)$ to the plane $2x + y - 2z = 10$.

11. Find the distance between the planes $x - 2y + 3z = 1$ and $x - 2y + 3z = 5$.

5.3 VECTOR PRODUCTS

In the last section we considered the scalar product of two vectors and discussed some physical and geometrical applications. When moments or angular momentum problems are considered, the resultant of two vectors is another vector perpendicular to the first two, which leads to the definition of the vector product:

$$C = A \times B = |A||B|(\sin \theta)u, \tag{1}$$

where θ is the angle between A and B and u is a unit vector perpendicular to A and B such that A, B, and C form a right-handed system as defined in Section 3.1 and shown in Figure 1. The vectors u and C clearly have the same direction since the angle θ between two vectors must satisfy the relationship $0 < \theta < \pi$. For the special cases $\theta = 0$ or π we have that $\sin \theta = 0$ and hence $A \times B = 0$. Thus the cross product of two vectors in the same or opposite directions is zero. This, of course, is consistent with the physical interpretation as a moment.

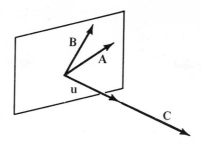

Figure 1.

As with the scalar product, the vector (or <u>cross</u>) product is usually computed using the component form:

$$A \times B = (a_2 b_3 - a_3 b_2)i + (a_3 b_1 - a_1 b_3)j + (a_1 b_2 - a_2 b_1)k \tag{2}$$

$$= \begin{vmatrix} i & j & k \\ a_1 & a_2 & a_3 \\ b_1 & b_2 & b_3 \end{vmatrix}. \tag{3}$$

Example 1.   Find the vector product of $A = i + j + k$ and
            $B = 2i - j + 3k$.

From Eq. (3) we have

$$A \times B = \begin{vmatrix} i & j & k \\ 1 & 1 & 1 \\ 2 & -1 & 3 \end{vmatrix} = 4i - j - 3k.$$

The answer in Example 1. can be partially checked by showing that it is perpendicular to both of the given vectors.

A geometrical interpretation of the cross product is obtained from Figure 2. The area of the parallelogram with sides parallel to A and B is $|A| \, |B| \sin \theta$, which is just the magnitude of $A \times B$ from Eq. (1). For a physical application of the

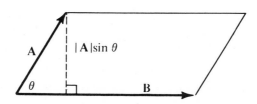

Figure 2.

vector product, let R denote the position vector of the point of application of a
force F, as shown in Figure 3.   Then the vector moment about an axis through 0

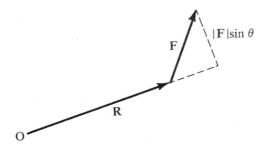

Figure 3.

perpendicular to the plane of R and F is given by

$$M = R \times F .$$ (4)

Note that M is a vector perpendicular to the plane of R and F in the right-handed
sense.   The scalar moment is simply the magnitude of $|M|$

$$|M| = |R| \, |F| \, \sin \theta ,$$ (5)

where $|F| \sin \theta$ represents the component of force acting perpendicular to R.
     The following properties of vector products can be verified using either Eq. (1)
or Eq. (3):

$$A \times B = - B \times A$$ (6)

$$A \times (B + C) = A \times B + A \times C$$ (7)

$$m (A \times B) = (mA \times B) = A \times (mB).$$ (8)

     Products of three vectors can also be found if they are multiplied in a mean-
ingful way.   The triple scalar product A · (B × C) is defined since B × C is a vector
which is then dotted with A.   If B × C is written as in the determinant form of
Eq. (3), then the triple scalar product may be computed as

$$A \cdot B \times C = \begin{vmatrix} a_1 & a_2 & a_3 \\ b_1 & b_2 & b_3 \\ c_1 & c_2 & c_3 \end{vmatrix}$$ (9)

since the dot product is simply the i component of A times the i component of B × C etc. Using cofactors then, the above determinant is precisely A · B × C.

Example 2.   Find A · B × C if A = i + j, B = i + j + k
                        and C = 2i − k.

From Eq. (9) we have

$$A \cdot (B \times C) = \begin{vmatrix} 1 & 1 & 0 \\ 1 & 1 & 1 \\ 2 & 0 & -1 \end{vmatrix} = 2.$$

Using properties of determinants, various rearrangements of the triple scalar product may be related to A · B × C:

$$A \cdot B \times C = A \times B \cdot C \tag{10}$$

$$= - B \cdot A \times C \tag{11}$$

plus others.

The geometrical interpretation of A · B × C is that the absolute value of A · B × C represents the volume of the three dimensional parallelopiped with sides parallel to A, B, and C , as illustrated in Figure 4. This can be reasoned as follows:

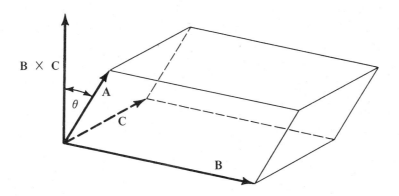

Figure 4.

using the definition of the dot product we have

$$A \cdot B \times C = |A| \ |B \times C| \ \cos \theta , \tag{12}$$

where θ is the angle between A and the vector B × C. Now |B × C| is the area of the base parallelogram and |A|cos θ is the "height" of the parallelopiped. If $0 \le \theta < \pi/2$, then A, B and C form a right-handed system and Eq. (12) yields a positive value.   If $\pi/2 < \theta \leq \pi$, then A, B and C form a left-handed system and cos θ is negative.   Thus the absolute value of A · B × C represents the volume of the parallelopiped.

Triple vector products are much more complicated.   For one thing the product A × B × C is not well defined without the use of parentheses.   If parentheses are inserted, there are two possible combinations of triple vector products which may be calculated from:

$$A \times (B \times C) = (A \cdot C)\ B - (A \cdot B)\ C \qquad\qquad (13)$$

and

$$(A \times B) \times C = (A \cdot C)\ B - (B \cdot C)\ A. \qquad\qquad (14)$$

From Eqs. (13) and (14) we easily see that

$$A \times (B \times C) \neq (A \times B) \times C, \qquad\qquad (15)$$

that is, the associative law does not hold.   Equations (13) and (14) will not be proved here, but their plausibility can be justified from a knowledge of the cross product.   As can be seen in Figure 5, if A × (B × C) is not the zero vector, then it

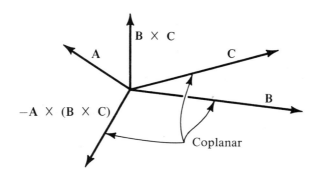

Figure 5.

must be perpendicular to B × C.   But B × C is perpendicular to both B and C and thus A × (B × C) must be in the plane of B and C and hence must be a linear combination of B and C, which is what is stated by Eq. (13).   A similar argument can be given for Eq. (14).

Example 3.   If A = i + j,   B = 2i + 3j − k   and   C = 4j + 3k, find A × (B × C).

Using Eq. (13) we have

$$A \times (B \times C) = (A \cdot C)\ B - (A \cdot B)\ C$$

$$= 4(2i + 3j - k) - 5(4j + 3k)$$

$$= 8i - 8j - 19k.$$

The student should rework Example 3. using Eq. (3) twice to verify the use of Eq. (13) and also compute (A × B) × C to verify Eq. (15).

The concept of a linear combination used in the above argument is illustrated in Figure 6, where as long as A and B are not parallel then any vector C in the plane of A and B can be written as a linear combination of A and B:

$$C = sA + tB, \tag{16}$$

for the appropriate choice of s and t.

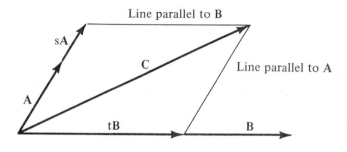

Figure 6.

Exercises

1. Let A = i + 2j + k, B = 3i - j and C = 4k - j.

    a) Find A · B × C   b) Find A × C

    c) Find A × (B × C)   d) Find (A × B) × C

    e) Find a unit vector perpendicular to A and B.

2. Find the equation of the plane passing through (1,2,-1) parallel to the plane of the vectors A = i + 2j - k and B = i + 2k.

3. Find the equation of the line through (-1,5,4) parallel to the line of intersection of the planes x - y + 3z = 10 and 2x + y - 4z = 5.

4. a) Find the area of the triangle with vertices at (3,1,2), (2,-3,4) and (1,5,4).
    b) Find the equation of the plane determined by the points of part a).

5. Repeat Exercise 4. with the three points $(0,2,0)$, $(-1,0,0)$ and $(0,0,5)$.

6. If $A = i + j$, $B = 2i - j$, and $C = -i + 6j$ find s and t so that $C = sA + tB$ and illustrate geometrically.

7. If $C = sA + tB$, evaluate the 3×3 determinant that has the components of A as the first row, the components of B as the second row and the components of C as the third row.

8. A force of 10 lbs. is applied in the positive x direction at the point $(1,1,3)$. Find the scalar moment of this force about the line $x = t$, $y = t$ and $z = t$.

9. Derive the identity

$$(A \times B) \times (C \times D) = (A \cdot B \times D) C - (A \cdot B \times C) D.$$

Can the left hand product also be written as a linear combination of A and B rather than C and D? If so find it.

10. Derive the identity

$$(A \times B) \cdot (C \times D) = (A \cdot C)(B \cdot D) - (A \cdot D)(B \cdot C).$$

5.4 VECTOR FUNCTIONS OF A SINGLE VARIABLE

A vector whose components depend on a variable t is called a vector-valued function and is denoted:

$$F(t) = a_1(t)i + a_2(t)j + a_3(t)k. \tag{1}$$

Since each of the components of $F(t)$ is a scalar function of a single variable, the theory of vector functions closely resembles that of real valued functions. For example, the vector function $F(t)$ has a limit at t_0 or is continuous at t_0 provided each of its components has a limit or is continuous at t_0.

Example 1.   If

$$F(t) = \sin(2t)i + \cos(2t)j + k$$

then

$$\lim_{t \to \frac{\pi}{4}} F(t) = i + k.$$

We also see that $F(t)$ is continuous for all t since each of its componets is continuous for all t.

The definition of the derivative of a vector function will be needed later and hence is given here for reference:

$$F'(t) = \frac{dF}{dt} = \lim_{\Delta t \to 0} \frac{F(t+\Delta t) - F(t)}{\Delta t} , \qquad (2)$$

if the limit exists. Since the difference of two vectors is computed by subtracting the components, Eq. (2) implies that

$$F'(t) = a_1'(t)i + a_2'(t)j + a_3'(t)k, \qquad (3)$$

provided each of the components has a derivative.

Example 2.   If

$$F(t) = \sin(2t)i + \cos(2t)j + k$$

then

$$F'(t) = 2\cos(2t)i - 2\sin(2t)j.$$

Using Eq. (3) the following properties of the derivative of vector functions can be verified:

$$\frac{d}{dt} (F + G) = F' + G' \qquad (4)$$

$$\frac{d}{dt} (sF) = s'F + sF' \qquad (5)$$

$$\frac{d}{dt} (F \cdot G) = F' \cdot G + F \cdot G' \qquad (6)$$

$$\frac{d}{dt} (F \times G) = F' \times G + F \times G' \qquad (7)$$

where the order of the cross products in Eq. (7) must be preserved.

Example 3.   Show that F' is perpendicular to F whenever F has
                          a constant magnitude.

From Eq. (12) of Section 5.2 we have

$$F \cdot F \;=\; |F|^2 \;=\; \text{constant},$$

since we know that $|F|$ is a constant.   Thus

$$\frac{d}{dt} (F \cdot F) = 0,$$

and using Eq. (6) we obtain

$$F' \cdot F + F \cdot F' \;=\; 0.$$

Since the dot product is commutative, this last equation yields

$$2F \cdot F' \;=\; 0, \tag{8}$$

which says that F' is either zero or perpendicular to F.

The vector $F(t)$ of Example 2. satisfies the conditions of Example 3.   The result
of Example 3. is very important in application problems involving circular or
spherical motion.

If the position vector that we used in previous sections is a function of t
then the tip of the vector describes a curve in three dimensional space as t varies:

$$R(t) = x(t)i + y(t)j + z(t)k. \tag{9}$$

We have actually already seen an example of a space curve when we found the vector
equation of a straight line:

$$R(t) = R_0 + t(R_1 - R_0), \tag{10}$$

where

$$x(t) = x_0 + t(x_1 - x_0),$$

$$y(t) = y_0 + t(y_1 - y_0),$$

$$z(t) = z_0 + t(z_1 - z_0).$$

Another position vector is given in Example 2, where a circle of unit radius in the
$z = 1$ plane is described.

Example 4.   The position vector

$$R(t) = \cos(t)i + \sin(t)j + tk$$

describes a circular helix, shown in Figure 1, since
$x(t) = \cos t$ and $y(t) = \sin t$ satisfy the relationship $x^2 + y^2 = 1$, while $z(t) = t$ increases continuously
with t.

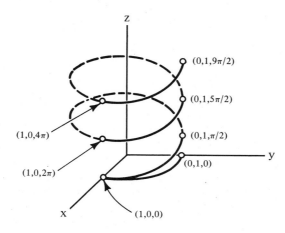

Figure 1.

The derivative of a position vector plays an important role in many applications.
Geometrically, the derivative of R(t) is tangent to the curve with vector equation
R(t). This can be illustrated by Figure 2, which shows the case for two dimensions.

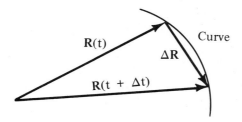

Figure 2.

Forming the difference quotient we have

$$\frac{\Delta R}{\Delta t} = \frac{R(t+\Delta t) - R(t)}{\Delta t} \quad , \tag{11}$$

and as $\Delta t \to 0$, R(t+Δt) approaches R(t) and ΔR approaches the tangent to the <u>smooth</u>
curve. From Eq. (2), as Δt approaches zero Eq. (11) becomes the defining equation

for $\dfrac{dR}{dt}$. The vector T:

$$T = \frac{x' \, i + y' \, j + z' \, k}{\sqrt{(x')^2 + (y')^2 + (z')^2}} \tag{12}$$

$$= \frac{R'}{|R'|} \tag{13}$$

is then a unit vector tangent to the curve described by R.

Example 5.   Find a unit vector tangent to the curve given by

$$R(t) = \cos(2t)i + \sin(2t)j + k.$$

The derivative is given by

$$R'(t) = -2\sin(2t)i + 2\cos(2t)j,$$

so that

$$|R'(t)| = \sqrt{4\sin^2 2t + 4\cos^2 2t} = 2.$$

Thus the desired unit vector is

$$T = -\sin(2t)i + \cos(2t)j.$$

Since the unit tangent vector has a constant magnitude, we may then conclude, by Example 3, that the vector T' must be perpendicular to the curve given by $R(t)$. A unit vector in the same direction as T' , and denoted by $N,$ is called the <u>principal normal</u>. For a space curve there are an infinite number of unit normals, all in the plane perpendicular to the curve. The principal normal is the one that is directed toward the center of curvature for a particular point.

Physically, the derivative $\dfrac{dR}{dt}$ represents the velocity when $R(t)$ represents the position of a body at any time t. This can be justified by a slight extension of Figure 2. as shown in Figure 3. Here ΔR represents the displacement during the time interval Δt. Thus $\Delta R/\Delta t$ represents the average velocity during the time interval Δt, and as Δt approaches zero, we obtain the instantaneous velocity, $\dfrac{dR}{dt}$. By a similar argument, $\dfrac{d^2 R}{dt^2}$ represents the acceleration of a body or particle with position vector $R(t)$. Since $R' = |R'|T$ we have

$$R'' = \frac{d}{dt}|R'|\ T + |R'|\frac{dT}{dt}$$

$$= |R'|'T + |R'||T'|N, \tag{14}$$

and thus the acceleration vector lies in the plane of the unit tangent and unit principal normal vectors.

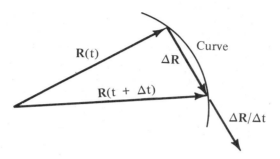

Figure 3.

The actual distance a particle moves in a time interval $[t_1, t_2]$ when its position is given by $R(t)$ is:

$$s = \int_{t_1}^{t_2}|R'|\,dt, \tag{15}$$

which is also frequently called the length of arc from t_1 to t_2. Equation (15) can be derived by using the basic definition of the integral as a limit of a sum and the pythagorean theorem. Consider Figure 4. where the length of ΔR_1, $|\Delta R_1|$, represents

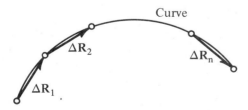

Figure 4.

an approximate length of arc over that subdivision. For n subdivisions then,

$$s = \sum_{i=1}^{n}|\Delta R_i|$$

$$= \sum_{i=1}^{n}\frac{|\Delta R_i|}{\Delta t_i}\Delta t_i \tag{16}$$

so as $\Delta t_i \to 0$ (or $n \to \infty$), we obtain:

$$s = \int_{t_1}^{t_2} \left| \frac{dR}{dt} \right| dt,$$

which is the same as Eq. (15).

Example 6.   Find the arc-length of the curve given by

$$R(t) = \cos(t)i + \sin(t)j + tk$$

from the point $(1,0,0)$ to the point $(1,0,2\pi)$.

To find the appropriate limits to use in Eq. (15), we must find the t values that when substituted in R yield the given points. In this case $t = 0$ and $t = 2\pi$ give the desired values and thus

$$s = \int_0^{2\pi} \left| -\sin(t)i + \cos(t)j + k \right| dt$$

$$= \int_0^{2\pi} \sqrt{\sin^2 t + \cos^2 t + 1} \; dt$$

$$= 2\pi\sqrt{2} \; .$$

In many applications an arc-length parameter is used as the variable rather than a variable such as t. The arc-length parameter $s(t)$ is defined by

$$s(t) = \int_{t_1}^{t} \left| \frac{dR}{dt} \right| dt \qquad t \geq t_1, \qquad (17)$$

where t is considered a variable upper limit. Note that $s(t)$ is always positive unless $t = t_1$, in which case it is zero. From the fundamental theorem of calculus we know that the derivative of Eq. (17) yields

$$\frac{ds}{dt} = \left| \frac{dR}{dt} \right| \qquad\qquad (18)$$

and hence

$$\frac{ds}{dt} = \sqrt{(x')^2 + (y')^2 + (z')^2} \ . \tag{19}$$

Since all the terms on the right of Eq. (19) can't be zero simultaneously, when R describes a smooth curve, we have that $\frac{ds}{dt} \neq 0$. Thus s(t) never reaches a relative maximum value, as shown in Figure 5, and we can conclude that for each   s   there is

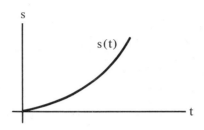

Figure 5.

only one   t.   This means that Eq. (17) can be solved for   t   as a function of s, t(s), and therefore   R   and its derivatives can be written in terms of the arc-length parameter s.

Example 7.   Rewrite the vector function

$$R(t) = \cos(2t)i + \sin(2t)j$$

in terms of its arc-length parameter.

We first of all must calculate s(t) as follows:

$$R'(t) = -2\sin(2t)i + 2\cos(2t)j$$

and therefore

$$s(t) = \int_0^t |R'| dt$$

$$= \int_0^t 2dt = 2t.$$

We have taken $t_1 = 0$ arbitrarily here, although this is the usual starting point. Solving for   t   in terms of   s   we get t = s/2, and

$$R(s) = \cos(s)i + \sin(s)j$$

gives R in terms of the arc-length parameter s.

In Example 7, $R(t)$ represents a point moving around a unit circle. At $t = 0$ the point starts at $(1,0)$ and thus $s(t)$ represents the length of the arc of a circle.

Exercises

1. A particle moves along the plane curve given by

$$R(t) = a \cos(\omega t)i + a \sin(\omega t)j.$$

   a) Find $R'(t)$, $R''(t)$, $|R'(t)|$ and $|R''(t)|$.

   b) Sketch the curve.

   c) What is the geometric relationship of R, R' and R''? Show them on the sketch.

   d) Find R in terms of the arc-length parameter.

2. Find $\dfrac{d}{dt} (R \times \dfrac{dR}{dt})$.

3. As t varies from -1 to 1, $R(t) = |t|i + t^2 j + k$ traces a curve. At what point on this curve is there no tangent?

4. Let $R(t) = e^t \cos(t)i + e^t \sin(t)j$.

   a) Sketch the curve for $0 \le t \le 2\pi$.

   b) Find a unit vector T tangent to the curve.

   c) Find T' and show that T and T' are perpendicular.

   d) Find R in terms of the arc-length parameter.

   e) Find $\dfrac{dR}{ds}$ and show that $\left|\dfrac{dR}{ds}\right| = 1$.

5. Show that $\dfrac{d}{dt} (F \times F' \cdot F'') = F \times F' \cdot F'''$.

6. A particle moves along a curve given by the position vector

$R(t) = 3t\,i + 2t^2\,j$.

a) Find the velocity and acceleration as a function of t.

b) Find the unit tangent vector T.

c) Find the normal vector dT/dt.

d) Write the acceleration vector as a linear combination of the unit tangent
vector and the normal vector found in part c).

7. Show that Eq. (14) can be written as

$$R'' = s''T + (s')^2\,kN,$$

where

$$k = |dT/ds|.$$

8. Show that the principal normal vector to the circular helix of Example 4. is
parallel to the xy plane and always directed towards the z axis.

9. Describe the space curve given by

$$R(t) = 2\cos(3t)i + \sin(3t)j + e^t k.$$

5.5 VECTOR FUNCTIONS AND DIRECTIONAL DERIVATIVES

In the preceding four sections on vectors we have used a coordinate system
which is fixed and therefore the unit vectors i, j and k have been constant. However,
some applications require the introduction of a coordinate system that moves with
a particular body or particle. A very good example of such a system is found when
motion in the xy plane is described in terms of the r, θ coordinates (Section 4.2),
as shown in Figure 1.

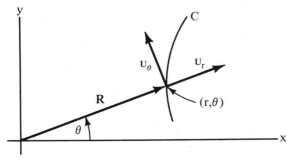

Figure 1.

In order to work with polar coordinates it is advantageous to use the unit vectors u_r and u_θ, also shown in Figure 1, where u_r is a unit vector in the direction of the position vector R and u_θ is a unit vector perpendicular to u_r in the direction of increasing θ. The vectors u_r and u_θ are related to i and j by:

$$u_r = \cos(\theta)i + \sin(\theta)j$$

$$(1)$$

$$u_\theta = -\sin(\theta)i + \cos(\theta)j,$$

where the basic definitions of the sine and cosine functions have been used. Equations (1) show that both u_r and u_θ vary in direction as the position vector $R(t)$ moves along the path C, since θ will vary from point to point. From this point of view then, the u_r and u_θ coordinate system is moving. Thus we must be careful in differentiating vectors written in terms of u_r and u_θ. From Eq. (1) we have

$$\frac{du_r}{d\theta} = u_\theta \quad \text{and} \quad \frac{du_\theta}{d\theta} = -u_r . \qquad (2)$$

The position vector R in terms of polar coordinates will be given by

$$R = r \, u_r \qquad (3)$$

where r and u_r will be functions of t if R is a function of t. Thus:

$$v = \frac{dR}{dt} = \frac{dr}{dt} u_r + r \frac{du_r}{dt} \qquad (4)$$

$$= r' \, u_r + r\theta' \, u_\theta . \qquad (5)$$

The chain rule and the first equation in (2) have been used to obtain the second term. In Eq. (5), r' is called the _radial_ component of the velocity and $r\theta'$ is the _transverse_ component.

Example 1.   The position of a person on a platform rotating with constant angular velocity 2 rad/sec is given by

$$R(t) = (1 + 3t) \, u_r . \qquad (6)$$

          Find the velocity at $t = 2$.

Calculating the velocity directly from Eq. (6) we have

$$v = \frac{dR}{dt} = 3u_r + (1 + 3t) \frac{du_r}{d\theta} \frac{d\theta}{dt}$$

$$= 3u_r + (2 + 6t) u_\theta \qquad (7)$$

since $d\theta/dt$ is the angular velocity. Thus at $t = 2$ we get

$$v(2) = 3u_r + 14 u_\theta.$$

The acceleration in terms of polar coordinates may be found by differentiating Eq. (5) and using Eqs. (2) appropriately. The result is given here for reference:

$$a = \frac{dv}{dt} = \left(\frac{d^2r}{dt^2} - r(\frac{d\theta}{dt})^2 \right) u_r + \left(r \frac{d^2\theta}{dt^2} + 2 \frac{dr}{dt} \frac{d\theta}{dt} \right) u_\theta \qquad (8)$$

where the first and third terms have their counterparts in Eq. (5). The second term, $-r(\theta')^2$, is a centripetal acceleration and the fourth term, $2r'\theta'$, is called the _Coriolis_ _acceleration_ , which is due partly to the change in the direction of the radial component of velocity and partly to the fact that as   r   changes the transverse component of velocity changes, as can be seen in Example 1. and Eq. (4).

Example 2.   Find the acceleration for the position vector of
                        Example 1.
     Using the velocity from Eq. (7) we have:

$$a = \frac{dv}{dt} = 3 \frac{du_r}{dt} + 6u_\theta + (2 + 6t) \frac{du_\theta}{dt}$$

$$= 3 \frac{d\theta}{dt} u_\theta + 6u_\theta + (2 + 6t) - u_r \frac{d\theta}{dt}$$

$$= -(4 + 12t) u_r + 12 u_\theta.$$

In Example 2, since $\frac{d\theta}{dt} = 2$ and $\frac{d^2\theta}{dt^2} = 0$, the total transverse component of a

is the Coriolis acceleration.

     The directional derivative of a function of several variables is a generalization of the partial derivatives that we considered in Chapter 3. The concept of the directional derivative is most easily understood by considering the temperature $T(x,y,z)$ throughout a room (pressure or density would work just as well). If a

thermometer is moved from one point to another, a change in temperature will be
noted, the amount of change depending on which direction is chosen. The directional
derivative, then, gives the rate of change in a specified direction. We will derive
the directional derivative for functions of three variables, but out of necessity
our illustrations are for functions of two variables, as shown in Figure 2.

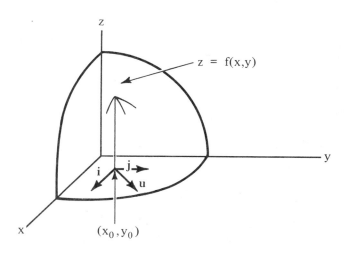

Figure 2.

The directional derivative of a function $f(x,y,z)$ in the direction of the unit
vector u is found by imagining a line passing through $(x_0,\ y_0,\ z_0)$ parallel to u.
If s denotes the displacement along the line in the direction of u, with $s = 0$
corresponding to $(x_0,\ y_0,\ z_0)$, then $\frac{df}{ds}$ at $s = 0$ is called the directional derivative
of f at $(x_0,\ y_0,\ z_0)$ in the direction of u. The directional derivative will depend
on the point $(x_0,\ y_0,\ z_0)$ and the specified direction. If $u = i$, then $\frac{df}{ds} = \frac{\partial f}{\partial x}$
and if $u = j$ then $\frac{df}{ds} = \frac{\partial f}{\partial y}$.

To calculate the directional derivative, $\frac{df}{ds}$, in any direction u, all we have
to do is apply the chain rule (Section 3.2) to $f(x,y,z)$ when x,y and z are functions
of the arc-length parameter s:

$$\frac{df}{ds} = \frac{\partial f}{\partial x} \frac{dx}{ds} + \frac{\partial f}{\partial y} \frac{dy}{ds} + \frac{\partial f}{\partial z} \frac{dz}{ds} \ . \tag{9}$$

Equation (9) may be rewritten as a dot product of two vectors. To do this, we
define

$$\text{grad } f = \nabla f = \frac{\partial f}{\partial x} i + \frac{\partial f}{\partial y} j + \frac{\partial f}{\partial z} k \tag{10}$$

and

$$u = \frac{dx}{ds} i + \frac{dy}{ds} j + \frac{dz}{ds} k . \tag{11}$$

Equation (10) is the definition of the <u>gradient</u> of a scalar function f and Eq. (11) we recognize from a previous exercise as a unit vector in the specified direction since s is the arc-length parameter and $R = x(s)i + y(s)j + z(s)k$ is the position vector of a point on the line through (x_0, y_0, z_0) parallel to u. Thus

$$\frac{df}{ds} = \nabla f \cdot u \tag{12}$$

is the desired representation of the directional derivative as a dot product. It is important to remember that grad f or ∇f denote vector quantities.

Example 3.   Find grad f if

$$f(x,y,z) = yz + xy + 2z^3.$$

From Eq. (10) we have

$$\nabla f = yi + (x + z)j + (y + 6z^2)k.$$

Example 4.   Find the directional derivative at $(1,2,1)$ of

$$f(x,y,z) = 4x^2 - y^2 + 2z$$

in the direction of the vector $2i + j + k$.

To use Eq. (12) we first find grad f as

$$\nabla f = 8xi - 2yj + 2k \Big|_{(1,2,1)} = 8i - 4j + 2k.$$

The unit vector u in the specified direction is given by

$$u = \frac{2i + j + k}{\sqrt{4 + 1 + 1}} = \frac{2i + j + k}{\sqrt{6}} .$$

Thus the desired directional derivative is given by

$$\frac{df}{ds} = \nabla f \cdot u = \frac{16 - 4 + 2}{\sqrt{6}} = \frac{7\sqrt{6}}{3} .$$

Example 5.   What is the maximum possible rate of change of the function

$$f = 4xy + 2yz^2 \text{ at } (1,1,0)?$$

To find the maximum rate of change of f, we find the directional deriva-

tive of f at the given point for any direction:

$$\frac{df}{ds} = \nabla f \cdot u = |\nabla f| \, |u| \cos \theta,$$

where θ is the angle between u and ∇f (which is fixed for the given point).
Thus

$$\frac{df}{ds} = |\nabla f| \cos \theta,$$

since |u| = 1. Hence the maximum rate of change of f is |∇f| and

$$|\nabla f| = \left. \left| 4y\hat{i} + (4x + 2z^2)\hat{j} + 4yz\hat{k} \right| \right|_{(1,1,0)}$$

$$= |4\hat{i} + 4\hat{j} + 0\hat{k}|$$

$$= \sqrt{32} .$$

The above examples have illustrated the computational aspects of Eqs. (10), (11)
and (12). In the next section we will pursue further the geometric interpretation
of the directional derivative and the gradient. It should be mentioned that the
gradient vector is an example of a vector field, a vector associated with each point
in space.

Exercises

1. A wheel of radius 3 inches is rotating with a constant angular speed of 0.5
    revolutions per second. Find the velocity and acceleration vectors in terms
    of u_r and u_θ for a point on the outer circumference. What are the magnitudes
    of the velocity and acceleration?

2. A mass moves in a circle of radius 4 inches with an angular displacement
    given by $\theta = 2 + t^2$. Find the velocity and acceleration vectors in terms of
    u_r and u_θ. What are the centripetal and transverse accelerations as a func-
    tion of time?

3. For $R = r(t) \, u_r(\theta)$ find v and a if $r = b(1 + \cos t)$ and $\theta = e^{-t}$.

4. Find the Coriolis acceleration for the planar motion given by $x(t) = t \cos 2t$ and $y(t) = t \sin 2t$. (Hint: Find $r(t)$ from the given terms and use the polar representation).

5. Find the directional derivative of $f = 2xz^4 - x^2y + z$ at the point $(-2,2,-1)$ in the direction of $2i - j - 2k$.

6. Find the directional derivative of $f = 2x^2y - 3yz^2$ at the point $(1,2,1)$ in the direction toward the point $(1,0,3)$. What are the maximum and minimum values of the directional derivative?

7. Show that the vector u of Eq. (11) can be written as

$$u = \cos(\alpha)i + \cos(\beta)j + \cos(\gamma)k,$$

where α, β and γ have been defined in Section 5.1. Use this form of u in Eq. (12) to rewrite Eq. (9) in terms of α, β and γ. (Hint: To find u as above, find the position vector for any point on the line parallel to u through (x_0, y_0, z_0) and then use the results of Section 5.4).

8. For a function of two variables show that u of Exercise 7. reduces to

$$u = \cos(\alpha)i + \sin(\alpha)j,$$

where α is the angle between the positive x axis and the specified direction. In this case show that the directional derivative is given by

$$\frac{df}{ds} = \frac{\partial f}{\partial x} \cos \alpha + \frac{\partial f}{\partial y} \sin \alpha.$$

(Hint: See Figure 2.)

9. Find the directional derivative of $f = xy^2 + 2\sin z$ at the point $(0,1,0)$ in the direction tangent to the curve $x = t$, $y = \cos t$ and $z = \sin t$.

10. Compute ∇f for each of the following functions.

a) $f = \sin xy + ze^x$

b) $f = x^2yz + 4xz^2$

c) $f = 2xye^{yz} + x^2 \ln y$

d) $f = g(r)$, $r = \sqrt{x^2 + y^2 + z^2}$

11. Find all functions f such that $\nabla f = zi + 2yj + (x + 1)k$.

12. Find all functions f such that

$$\nabla f = (y^2 - 2xyz^3)i + (2xy - x^2z^3)j + (4z^3 - 3x^2y\ z^2)k$$

13. Show that there is no function $f(x,y,z)$ for which $\nabla f = zi + 2yj + k$.

5.6 THE GRADIENT AND ITS APPLICATIONS

In the last section the concept of the directional derivative was developed and the concept of the gradient of a scalar function was introduced. There are some important consequences of the gradient that we would like to discuss here. In the previous work we saw that the derivative of f in the direction of the unit vector u was given by

$$\frac{df}{ds} = \nabla f \cdot u = |\nabla f|\cos\theta, \tag{1}$$

where θ is the angle between ∇f and u, and ∇f was defined as

$$\nabla f = \frac{\partial f}{\partial x} i + \frac{\partial f}{\partial y} j + \frac{\partial f}{\partial z} k. \tag{2}$$

The first important result is that grad f is a vector in the direction of the maximum rate of increase of the function f. This follows from Eq. (1) since the maximum value of $\frac{df}{ds}$ is $|\nabla f|$ and this value is obtained when u and grad f have the same direction, so that $\theta = 0$.

Example 1.   If $f(x,y) = \sqrt{4 - x^2 - y^2}$,

then

$$\nabla f = -\frac{x}{\sqrt{4 - x^2 - y^2}} i - \frac{y}{\sqrt{4 - x^2 - y^2}} j.$$

Thus at the point (1,1)

$$\nabla f = -\frac{1}{\sqrt{2}} i - \frac{1}{\sqrt{2}} j,$$

which is a vector that "points" in the direction of maximum increase of $f(x,y)$ as shown in Figure 1.

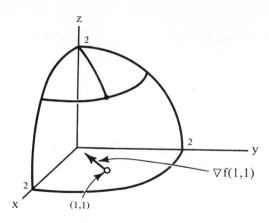

Figure 1.

Example 1. and Figure 1. illustrate the above concept for a function of two variables. Unfortunately it cannot be illustrated so nicely for a function of three variables, although the result still holds.

One of the many applications of the first result is in the numerical solution of either optimization problems or non-linear equations. In these problems the gradient vector will give the direction of the greatest rate of change of the function, and hence from a trial point successive increments can be made in the appropriate direction to yield the greatest increase (or decrease) in the function. From Figure 1. it is clear that once a maximum point (or a minimum point) is reached, then $\nabla f = 0\mathbf{i} + 0\mathbf{j}$ -- precisely the necessary condition for a relative maximum (or minimum) to exist, as we saw in Section 3.3.

The second important result concerns the equation $f(x,y,z) = C$. One can think of solving this equation for z as a function of x and y, and thus the equation represents a surface in three dimensional space. If the given surface passes through the point (x_0, y_0, z_0), then $f(x_0, y_0, z_0)$ must equal C and thus

$$f(x,y,z) = f(x_0, y_0, z_0) \qquad (3)$$

insures that the surface passes through (x_0, y_0, z_0). In this case grad f is perpendicular to the surface at the point (x_0, y_0, z_0). This can be reasoned with the aid of Figure 2. Let P be any path through (x_0, y_0, z_0) lying on the surface $f(x,y,z) = C$. Then, since f does not vary along the path,

$$\frac{df}{ds} = 0 = \nabla f \cdot \mathbf{u} \qquad (4)$$

where $\mathbf{u}$ is a unit vector tangent to the path and hence tangent to the surface $f(x,y,z) = C$. Since we may pick any path through (x_0, y_0, z_0) lying on the surface, $\mathbf{u}$ is an arbitrary unit vector tangent to the surface and Eq. (4) indicates that ∇f

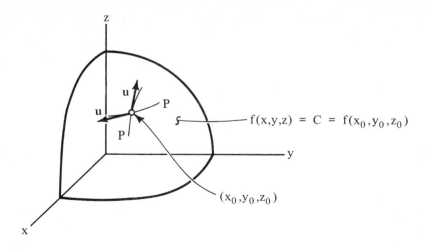

Figure 2.

must be perpendicular to the surface $f(x,y,z) = C$. (See Section 5.2 for a discussion of the dot product and perpendicularity.)

Example 2.   Find a vector perpendicular to the curve

$$f(x,y) = 4 - x^2 - y^2 \quad \text{at} \quad (1,1).$$

Using the above results we have that

$$\nabla f = -2\mathbf{i} - 2\mathbf{j}$$

is perpendicular to $4 - x^2 - y^2 = 2$ at $(1,1)$. Notice that the constant 2 is obtained by substituting $(1,1)$ into $f(x,y)$. The geometrical relation of the curve and ∇f are depicted in Figure 3. The "surface" in this case is really a curve in the xy plane, as the given function depends on only two variables and thus when $f(x,y)$ is set equal to a constant, an equation in the xy plane is obtained.

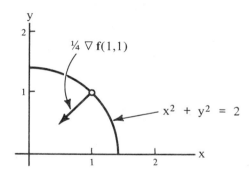

Figure 3.

Example 3.   Find a vector perpendicular to the surface

$$z = 6 - x^2 - y^2 \quad \text{at} \quad (2,1,1).$$

In this case $f(x,y,z)$ is given by

$$f(x,y,z) = x^2 + y^2 + z - 6 = 0 \qquad (5)$$

so that

$$\nabla f = 2x\mathbf{i} + 2y\mathbf{j} + \mathbf{k}$$

$$= 4\mathbf{i} + 2\mathbf{j} + \mathbf{k} \quad \text{at} \quad (2,1,1),$$

is perpendicular to the given surface. Notice that $(2,1,1)$ satisfies Eq. (5), so that the surface does pass through the given point.

Example 4.   Find a vector perpendicular to the sphere

$$x^2 + y^2 + z^2 = 4.$$

In this case we have that $f(x,y,z) = x^2 + y^2 + z^2$ and hence ∇f is given by

$$\nabla f = 2x\mathbf{i} + 2y\mathbf{j} + 2z\mathbf{k}. \qquad (6)$$

Equation (6) we recognize as a constant multiple of the position vector (Section 5.4) going from the origin to the point (x,y,z) lying on the given sphere. Since the sphere is centered at the origin, the vector in Eq. (6) must be perpendicular to the sphere for each point (x,y,z) lying on the sphere.

Example 5.   Find the equation of the plane tangent to the surface
$z = f(x,y)$ at (x_0,y_0).

Consider $F(x,y,z) = f(x,y) - z = 0$, then

$$\nabla f = \frac{\partial f}{\partial x}\mathbf{i} + \frac{\partial f}{\partial y}\mathbf{j} - \mathbf{k} \quad \text{at} \quad (x_0,y_0)$$

is perpendicular to the surface. Any vector in the desired plane passing through (x_0,y_0,z_0) is of the form

$$R - R_0 = (x - x_0)\mathbf{i} + (y - y_0)\mathbf{j} + (z - z_0)\mathbf{k},$$

which must be perpendicular to ∇f:

$$\nabla f \cdot (R - R_0) = 0.$$

Thus

$$\frac{\partial f}{\partial x}(x - x_0) + \frac{\partial f}{\partial y}(y - y_0) - (z - z_0) = 0, \tag{7}$$

is the desired tangent plane when f_x and f_y are evaluated at (x_0, y_0).

The tangent plane obtained in Example 5. is exactly the same one that was found in Section 3.1 by another method.

The method of Lagrange multipliers encountered in Section 3.3 can be derived using the gradient and the perpendicularity result studied above. The problem to be considered here is that of optimizing $f(x,y,z)$ subject to the condition $g(x,y,z) = 0$ and $h(x,y,z) = 0$. Referring to Figure 4, we see that $f(x,y,z)$ is to be optimized

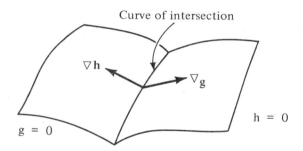

Figure 4.

while (x,y,z) varies along the curve of intersection of the two surfaces $g = 0$ and $h = 0$. Now we know that $\frac{df}{ds} = 0$ for an optimum point, which from Eq. (1) says ∇f must be perpendicular to u, the vector tangent to the curve. But we also know that ∇h and ∇g are perpendicular to each of the surfaces and hence are also perpendicular to the curve of intersection. Thus ∇f, ∇h and ∇g must be coplanar, and hence ∇f is a linear combination (Section 5.3) of ∇h and ∇g:

$$\nabla f + \lambda_1 \nabla h + \lambda_2 \nabla g = 0. \tag{8}$$

Equation (8) represents three scalar equations, which along with $h(x,y,z) = 0$ and $g(x,y,z) = 0$ are five equations for the five unknowns x,y,z,λ_1, and λ_2. You will recognize these five equations as precisely the ones we used previously, where λ_1 and λ_2 are the Lagrange multipliers.

The remaining sections of this chapter will be devoted to the development of the integral theorems of vector calculus. These results involve the concepts of the divergence and the curl of a vector field, which are introduced here since their calculation closely resembles the calculation of the gradient.

At this point we will give the mathematical calculation of the divergence and curl and defer the physical interpretation until the appropriate integral theorems have been introduced. To this end, then, we define the del operator as

$$\nabla = \frac{\partial}{\partial x} i + \frac{\partial}{\partial y} j + \frac{\partial}{\partial z} k, \tag{9}$$

which "looks" like a vector, but since it has no magnitude or direction it is not really a vector. However, it is used as a vector when multiplied times a scalar, or dotted, or crossed with a vector:

$$\nabla f = \left(\frac{\partial}{\partial x} i + \frac{\partial}{\partial y} j + \frac{\partial}{\partial z} k \right) f$$

$$= \frac{\partial f}{\partial x} i + \frac{\partial f}{\partial y} j + \frac{\partial f}{\partial z} k, \tag{10}$$

which we see is the gradient of f as defined previously. Likewise

$$\nabla \cdot F = \left(\frac{\partial}{\partial x} i + \frac{\partial}{\partial y} j + \frac{\partial}{\partial z} k \right) \cdot (F_1 i + F_2 j + F_3 k)$$

$$= \frac{\partial F_1}{\partial x} + \frac{\partial F_2}{\partial y} + \frac{\partial F_3}{\partial z} \tag{11}$$

is known as the divergence of F, and

$$\nabla \times F = \left(\frac{\partial}{\partial x} i + \frac{\partial}{\partial y} j + \frac{\partial}{\partial z} k \right) \times (F_1 i + F_2 j + F_3 k)$$

$$= \left(\frac{\partial F_3}{\partial y} - \frac{\partial F_2}{\partial z} \right) i + \left(\frac{\partial F_1}{\partial z} - \frac{\partial F_3}{\partial x} \right) j + \left(\frac{\partial F_2}{\partial x} - \frac{\partial F_1}{\partial y} \right) k \tag{12}$$

is known as the curl of F.

Example 6.   Calculate the divergence and the curl of the vector
                        field

$$F(x,y,z) = x^2 i + xyj + yzk.$$

Using Eq. (11) we find that

$$\nabla \cdot F = \frac{\partial}{\partial x} (x^2) + \frac{\partial}{\partial y} (xy) + \frac{\partial}{\partial z} (yz)$$

$$= 2x + x + y = 3x + y$$

is the divergence of F while Eq. (12) yields

$$\nabla \times F = \left(\frac{\partial (yz)}{\partial y} - \frac{\partial (xy)}{\partial z} \right) i + \left(\frac{\partial (x^2)}{\partial z} - \frac{\partial (yz)}{\partial x} \right) j$$

$$+ \left(\frac{\partial (xy)}{\partial x} - \frac{\partial (x^2)}{\partial y} \right) k$$

$$= zi + yk$$

as the curl of F.

Notice from Eqs. (11) and (12) and the example that the divergence of a vector field
is a scalar quantity and that the curl is another vector field. You will recall
from the end of Section 5.5 that a vector field is a vector associated with each
point in space.

Exercises

1. In what direction is the directional derivative of $f = x^2 + 2y^2 z$ a maximum
    at the point $(1,2,2)$?

2. What is the maximum rate of change of the function $f(x,y) = x^2 y + e^x \sin y$ at
    $(1, \pi/2)$ and in what direction does it occur?

3. Suppose that it is desired to find the maximum value of the function

$$f(x,y) = 4 - x^2 - y^2 + 2x + 2y$$

    using a numerical scheme. Evaluate f at $(0,0)$ and find the direction in

which f changes most rapidly at (0,0). Find a new point (x_1, y_1) which is
a unit distance from (0,0) in the direction just obtained and evaluate f
at (x_1, y_1) (the value should be larger than at (0,0)). Continuing in this
fashion the maximum point will eventually be reached (shorter steps will
have to be taken when the function values start decreasing at successive
points).

4. Let $f = 1 - 2x^2 - y^2$. Find ∇f at (1,1) and prove that ∇f is perpendicular
to the curve $f = -2$ passing through the point (1,1). Illustrate geometrically.

(Hint: Find parametric equations of the curve and use them to find a
tangent vector to the curve at the point (1,1).)

5. Suppose that you are at the point $(1,1,\sqrt{2})$ on the surface $z = \sqrt{4 - x^2 - y^2}$.
In what direction should you move

a) so that your rate of change will be zero?
b) so that your rate of increase will be the greatest?
c) so that your rate of decrease will be the greatest?
Draw a sketch to illustrate these results.

6. Find a unit vector perpendicular to each of the following surfaces.

a) $x^2 + z^2 = 5$ at (2,1)

b) $x^2 z + 2xy + yz^3 = 1$ at (1,2,-1)

c) $z = \sqrt{x^2 + y^2}$ at $(1,1,\sqrt{2})$

7. At what angle does the line $x = y = 2z$ intersect the ellipsoid $2x^2 + y^2 + z^2 = 13$?

8. Show that the unit vector perpendicular to the surface $z = f(x,y)$ is given by

$$N = \frac{\frac{\partial f}{\partial x} i + \frac{\partial f}{\partial y} j - k}{\sqrt{f_x^2 + f_y^2 + 1}} .$$

9. Derive the Lagrange multiplier equations for the problem of maximizing $f(x,y,z)$
subject to the constraint $g(x,y,z) = 0$. Give a discussion paralleling that of the
text.

10. Find the divergence and the curl of each of the following vector functions.

  a) $F = xy\mathbf{i} - y^2\mathbf{j} + 2xyz\mathbf{k}$

  b) $F = x\mathbf{i} + y\mathbf{j} + z\mathbf{k}$ ($= R$)

  c) $F = y^3\mathbf{j} + \sin(z)\mathbf{k}$

  d) $F = \sin(xy)\mathbf{i} + e^{xy}\mathbf{j} + y\cos(z^2)\mathbf{k}$

11. The __Laplacian__ is defined as $\nabla^2 = \nabla \cdot \nabla$.

  a) Find $\nabla \cdot \nabla f = \nabla^2 f$ when $f = x^2 y^3 z^{-1}$.

  b) Find $\nabla \cdot \nabla f = \nabla^2 f$ in terms of partial derivatives of f.

  c) The equation $\nabla^2 u = 0$ is known as Laplace's equation. Show that $u = \dfrac{1}{r}$,
      where r is the magnitude of the position vector, satisfies Laplace's
      equation.

12. Verify that the following vector identities are true.

  a) $\nabla \cdot \phi F = \phi \nabla \cdot F + F \cdot \nabla\phi$

  b) $\nabla \times \phi F = \phi \nabla \times F + \nabla\phi \times F$

  c) $\nabla \times (\nabla\phi) = 0$

  d) $\nabla \times (\nabla \times F) = 0$

5.7 LINE INTEGRALS, CONSERVATIVE FIELDS AND GREEN'S THEOREM

   The integral theorems to be introduced in Section 5.9 involve both line integrals and surface integrals. In this section we will define line integrals and give some results involving them and in Section 5.8 surface integrals will be discussed.

   Line integrals very frequently arise when various work problems are encountered, say in mechanical or electrical systems. In this connection then let $F(x,y,z)$ denote a force field throughout a three dimensional region. If the force F moves a particle or mass from point A to point B along a path described by the position vector R, then the work done will be given by

$$w = \int_A^B F \cdot dR . \tag{1}$$

Equation (1) is the <u>line integral</u> of F along the path R from A to B and is the generalization of the one dimensional integral $\int_a^b f(x)dx$, where $f(x)$ is the force required to move a mass from a to b along a straight line. The derivation of Eq. (1) follows steps similar to the development of the arc-length done in Section 5.4, where

$$\Delta w_i = F_i \cdot \Delta R_i \tag{2}$$

represents the increment of work in moving along a straight line segment approximation to the curve. These increments are then summed and when $|\Delta R_i| \to 0$, Eq. (1) is obtained.

The evaluation of Eq. (1) can be done in several ways. One way is to write the position vector in terms of a parameter, denoted as $R(t)$, which in turn defines $x(t)$, $y(t)$ and $z(t)$. If these latter functions are substituted in the vector F of Eq. (1), we then have a regular one-dimensional integral. Example 1. will illustrate this technique.

Example 1.

Evaluate $\int_{(0,0)}^{(1,1)} F \cdot dR$, when $F = xy i + x j$,

along the path $y = x^2$.

The desired parametric representation is given by

$$R(t) = t i + t^2 j$$

and hence $x(t) = t$ and $y(t) = t^2$. Also dR is computed as

$$dR = \frac{dR}{dt} dt = (i + 2t j)dt$$

and thus

$$\int_{(0,0)}^{(1,1)} F \cdot dR = \int_0^1 (t^3 i + t j) \cdot (i + 2t j)dt.$$

Notice that the vector $(t^3 i + tj)$ appearing in the right hand integral is the force vector evaluated along the given path, which is denoted here as $x(t) = t$ and $y(t) = t^2$. Carrying out the dot product we then obtain

$$\int_{(0,0)}^{(1,1)} F \cdot dR = \int_0^1 (t^3 + 2t^2) dt$$

$$= \frac{t^4}{4} + \frac{2t^3}{3} \Big|_0^1 = \frac{11}{12} .$$

The limits for the t integration in Example 1. are determined so that $R(t)$ is at $(0,0)$ when $t = 0$ and $R(t)$ is at $(1,1)$ when $t = 1$. It is clear from the above example that

$$\int_A^B F \cdot dR = -\int_B^A F \cdot dR, \qquad (3)$$

since what is involved is a reversal of the limits for the parameter t, and this changes the sign of the ordinary integral.

One important property of line integrals is that the value of the integral in general depends on the path chosen, as shown in the next example.

Example 2.   Evaluate the same integral as in Example 1. along the straight line path joining the points.

In this case $R(t)$ becomes

$$R(t) = ti + tj$$

so that $x(t) = t$, $y(t) = t$ and $dR = (i + j)dt$. Thus

$$\int_{(0,0)}^{(1,1)} F \cdot dR = \int_0^1 (t^2 i + tj) \cdot (i + j) dt$$

$$= \int_0^1 (t^2 + t) dt = \frac{5}{6} .$$

Comparing the results of Examples 1. and 2. shows that the path chosen can affect the value of a line integral.

In most problems it is possible to find more than one parametric representation. However, it can be shown that the parametric representation chosen does not affect the value of the line integral. Example 3. illustrates this point.

Example 3.

$$\text{Evaluate} \int_{(1,1,1)}^{(3,2,4)} F \cdot dR, \text{ when } F = xi - zj + yk,$$

along the straight line path joining the points.

One possible parametric representation is given by

$$R(t) = (1 + 2t)i + (1 + t)j + (1 + 3t)k,$$

in which case we have

$$\int_{(1,1,1)}^{(3,2,4)} F \cdot dR = \int_0^1 ((1 + 2t)i - (1 + 3t)j + (1 + t)k) \cdot (2i + j + 3k)dt$$

$$= \int_0^1 (4 + 4t)dt = 6.$$

A second possible parametric representation is given by

$$R(t) = (3 + 2t)i + (2 + t)j + (4 + 3t)k$$

in which case we get

$$\int_{(1,1,1)}^{(3,2,4)} F \cdot dR = \int_{-1}^0 ((3 + 2t)i - (4 + 3t)j + (2 + t)k) \cdot (2i + j + 3k)dt$$

$$= \int_{-1}^0 (8 + 4t)dt = 6.$$

Notice that when the parametric representation is changed, the limits will change accordingly.

A second method of evaluating line integrals is illustrated in Example 4, where the line integral of Eq. (1) is written in the form

$$\int_A^B F \cdot dR = \int_A^B F_1(x,y,z)dx + F_2(x,y,z)dy + F_3(x,y,z)dz \qquad (4)$$

since

$$dR = dx\ i + dy\ j + dz\ k. \qquad (5)$$

Example 4.

$$\text{Evaluate} \int_{(1,1)}^{(2,4)} F \cdot dR \text{ when } F = 2xy i + (x^2 + 1)j.$$

Writing the line integral in the form given in Eq. (4), we have

$$\int_{(1,1)}^{(2,4)} F \cdot dR = \int_{(1,1)}^{(2,4)} 2xy\ dx + (x^2 + 1)dy. \qquad (6)$$

Now, the integrand appearing in Eq. (6) (or in Eq. (4)) is of the form of a differential (Secton 3.1). For this example we recognize the integrand as the differential of the function $x^2y + y$. That is, if $f = x^2y + y$ then

$$df = 2xy\ dx + (x^2 + 1)dy.$$

Using this in Eq. (6) we obtain

$$\int_{(1,1)}^{(2,4)} F \cdot dR = \int_{(1,1)}^{(2,4)} df$$

$$= f(x,y) \Big|_{(1,1)}^{(2,4)} = 18.$$

This example, while illustrating an important method of evaluating line integrals, has raised some basic questions that need to be answered. It is quite evident that we needed no information concerning the path (as we will show, the integral of Example 4. is independent of which path is chosen) and it is not obvious how one recognizes that the integrand is an exact differential.

Both of the concepts just discussed are related to the idea of <u>conservative</u> vector fields. There are several ways one can define a conservative field -- all

equivalent to each other. From an applications point of view, a conservative
vector field F(x,y,z) has the property that the line integral $\int_C$ F · dR is zero
for any closed curve C. What this says is that energy is conserved, since no work
is done when you return to the starting position. This definition, however, is not
very practical for determining whether a given vector field is conservative or not
since it is not possible to calculate all such integrals. Thus we will define a
conservative vector field as one whose curl is zero. That is, F(x,y,z) is conservative if ∇ × F is the zero vector.

Example 5.   Show that $F = (2xy + z^3)i + x^2j + 3xz^2k$ is a
                 conservative field.

   To find the curl of F we use the determinant notation introduced in
   Section 5.3 Hence

$$\nabla \times F = \begin{vmatrix} i & j & k \\ \dfrac{\partial}{\partial x} & \dfrac{\partial}{\partial y} & \dfrac{\partial}{\partial z} \\ 2xy + z^3 & x^2 & 3xz^2 \end{vmatrix}$$

$$= 0i + (3z^2 - 3z^2)j + (2x - 2x)k = 0,$$

   which says that F is conservative by the above definition.

The reader can now verify that the vector field of Example 4. is also conservative.
   The geometrical and physical significance of conservative fields is brought
out by the following: if F is conservative, then it possesses all of the following
properties:

   1) F is the gradient of a scalar function (F = ∇φ).
   2) The line integral of F around any closed curve is zero.
   3) The line integral of F from a point Q to a point P is independent
       of the path.

The converse of these statements are also true. That is, if any one of the above
holds, then F is conservative. We will not prove these in a rigorous fashion,
but illustrate the concepts involved in proving the first result with Example 6.
The other results follow from the first and will be left as exercises.

Example 6.   Show that the vector function of Example 5. is the
gradient of a scalar function ϕ.

We already know that $F = (2xy + z^3)i + x^2j + 3xz^2k$ is conservative
from Example 5. To find ϕ so that $F = \nabla\phi$, we simply equate components:

$$F_1 = \frac{\partial\phi}{\partial x}, \quad F_2 = \frac{\partial\phi}{\partial y} \quad \text{and} \quad F_3 = \frac{\partial\phi}{\partial z}. \tag{7}$$

Using the first equation we have

$$\frac{\partial\phi}{\partial x} = F_1 = 2xy + z^3$$

so that

$$\phi(x,y,z) = x^2y + xz^3 + h(y,z). \tag{8}$$

The function $h(y,z)$ is the "constant" of integration. Using Eq. (8)
in the middle equation of (7) we obtain

$$x^2 + \frac{\partial h}{\partial y} = F_2 = x^2$$

and thus

$$\frac{\partial h}{\partial y} = 0 \quad \text{or} \quad h(y,z) = f(z), \tag{9}$$

where again $f(z)$ is the "constant" of integration. Putting Eq. (9)
into Eq. (8) and using the third equation of (7) we have finally

$$3xz^2 + \frac{df}{dz} = F_3 = 3xz^2$$

from which

$$\frac{df}{dz} = 0 \quad \text{or} \quad f(z) = \text{constant.} \tag{10}$$

Equations (8), (9) or (10) thus yield

$$\phi(x,y,z) = x^2y + xz^3 + \text{constant} \tag{11}$$

as the required scalar function. It can easily be shown that $\nabla\phi = F$
for the function given in Eq. (11).

The steps used in Example 6. are exactly the same ones that are used in deriving the first result for general conservative vector fields. In these cases, of course, the components of F are not known explicitly and thus integrals such as $\int F_1(x,y,z)dx$ are involved. The converse of statement 1) is that if $F = \nabla\phi$ then F is conservative. This has already been proved in Exercise 12c of Section 5.6, since $\nabla \times (\nabla\phi)$ is zero for any $\phi(x,y,z)$ and thus $F = \nabla\phi$ is conservative.

When F is not conservative, the same steps as shown in Example 6. would yield inconsistent statements in Eq. (9) or (10). Thus if an x appeared in Eq. (9) or an x or y appeared in Eq. (10), then one could not proceed further and no ϕ could be found such that $F = \nabla\phi$.

Example 7.   Show that no ϕ exists for which $F = \nabla\phi$
                   when $F = 2xy\mathbf{i} + x^3y\mathbf{j}$

If $F = \nabla\phi$, then the $\mathbf{i}$ and $\mathbf{j}$ compoents must be equal. Equating the $\mathbf{i}$ components we get

$$\frac{\partial\phi}{\partial x} = 2xy$$

or

$$\phi(x,y) = x^2y + h(y) \tag{12}$$

Equating the $\mathbf{j}$ components yields

$$\frac{\partial\phi}{\partial y} = x^3y$$

and using Eq. (12) we get

$$x^2 + \frac{dh}{dy} = x^3y$$

or

$$\frac{dh}{dy} = x^3y - x^2. \tag{13}$$

Equation (13) yields a contradiction, since we know from Eq. (12) that h must be a function of y only. Thus no ϕ exists for which $F = \nabla\phi$.

It can easily be checked that the vector field F of Example 7. is not conservative.

An important result is obtained for line integrals around closed paths lying in the xy plane. In this case we obtain Green's theorem, which can be written mathematically as

$$\int_C F_1(x,y)dx + F_2(x,y)dy = \iint_R \left(\frac{\partial F_2}{\partial x} - \frac{\partial F_1}{\partial y}\right)dA \qquad (14)$$

assuming the integrals involved exist. In Eq. (14) the curve C encloses the two dimensional region R and the direction of integration is such that the region is on the left as one proceeds along the curve. The derivation of Eq. (14) actually starts on the right hand side as follows:

$$\iint_R \frac{\partial F_2}{\partial x}\,dA = \int_d^e \int_{x_1(y)}^{x_2(y)} \frac{\partial F_2}{\partial x}\,dxdy, \qquad (15)$$

where the limits are obtained with the aid of Figure 1. The integration with respect

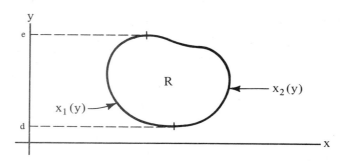

Figure 1.

to x is easily obtained so that

$$\iint_R \frac{\partial F_2}{\partial x}\,dA = \int_d^e F_2(x,y)\,\Big|_{x=x_1(y)}^{x=x_2(y)}\,dy$$

$$= \int_d^e \{F_2(x_2(y),y) - F_2(x_1(y),y)\}dy$$

$$= \int_d^e F_2(x_2(y),y)dy + \int_e^d F_2(x_1(y),y)dy$$

$$= \int_C F_2\,dy, \qquad (16)$$

since the sum of the last two integrals represents the indicated line integral. In
a similar fashion we can obtain

$$-\iint_R \frac{\partial F_1}{\partial y} \, dA = \int_C F_1 \, dx,$$ (17)

which when added to Eq. (16) yields Eq. (14).

Example 8.

$$\text{Evaluate} \int_C xy^2 \, dx + x^3 y \, dy \quad \text{around}$$

the triangle formed by $y = 0$, $x = 1$ and $y = x$.

Using Eq. (14) we obtain

$$\int_C xy^2 dx + x^3 y \, dy = \int_0^1 \int_0^x (3x^2 y - 2xy) \, dy \, dx$$

$$= \int_0^1 (3x^2 - 2x) \left. \frac{y^2}{2} \right|_0^x dx$$

$$= \frac{1}{2} \int_0^1 (3x^4 - 2x^3) \, dx = \frac{1}{20} .$$

An alternate form of Green's theorem results when Eq. (14) is rewritten in
vector form. The student may easily verify that Eq. (14) can be written as

$$\int_C \mathbf{F} \cdot d\mathbf{R} = \iint_R (\nabla \times \mathbf{F} \cdot \mathbf{k}) \, dA.$$ (18)

For conservative fields in the plane, Eq. (18) shows that the integral around any
closed path will be zero, since the right side is clearly zero. Stokes' Theorem,
to be discussed later, is a generalization of Eq. (18) for line integrals in three
dimensional space.

There are some very important vector fields for which Eq. (14) or Eq. (18)
appears to yield valid results, but which really are not valid. These cases are
illustrated in the following example.

Example 9.

$$\text{Evaluate} \int_C F \cdot dR \text{ counterclockwise around the circle}$$

of radius 1 centered at the origin when

$$F = \frac{-y\mathbf{i} + x\mathbf{j}}{x^2 + y^2}.$$

It is easily verified that $\nabla \times F$ is zero and hence Eq. (18) appears

to give $\int_C F \cdot dR = 0$. However, the vector F is not defined

at (0,0) and hence $\nabla \times F$ is really zero everywhere except at the origin, which is inside the given circle. Thus Green's Theorem is not applicable and the integral must be evaluated as follows

$$\int_C F \cdot dR = \int_0^{2\pi} \frac{-\sin(\theta)\mathbf{i} + \cos(\theta)\mathbf{j}}{\sin^2\theta + \cos^2\theta} \cdot (-\sin(\theta)\mathbf{i} + \cos(\theta)\mathbf{j})d\theta$$

since $x = \cos\theta$ and $y = \sin\theta$ is a parametric representation of the given circle. Thus

$$\int_C F \cdot dR = \int_0^{2\pi} (\sin^2\theta + \cos^2\theta)d\theta = 2\pi,$$

since $\sin^2\theta + \cos^2\theta = 1.$

Exercises 9. and 10. will illustrate an alternate way of evaluating such integrals when the paths are more complicated.

Exercises

1. Evaluate $\int_A^B F \cdot dR$ for each of the following.

a) $F = (x^2 + 1)\mathbf{i} + (y + x)\mathbf{j}$ along the x axis from the origin to x = 2 and then vertically along x = 2 to y = 1.

b) $F = (x^2 + 1)\mathbf{i} + (y + x)\mathbf{j}$ along the straight line segment joining (0,0) and (2,1).

c) $F = (x^2 + 1)i + (y + x)j$ along the curve $y = \frac{1}{2} x^2$ from $(0,0)$ to $(2,2)$.

d) $F = yi - xj$ along the top half of the circle $x^2 + y^2 = 4$ from $(-2,0)$ to $(2,0)$.

e) $F = yi - xj$ along the top half of the circle $x^2 + y^2 = 4$ from $(2,0)$ to $(-2,0)$.

f) $F = 2yi - 3xj + z^2k$ along the circular helix given by

$$R(t) = \cos(t)i + \sin(t)j + tk$$

from the point $(1,0,0)$ to the point $(1,0,2\pi)$.

2. Show that $\displaystyle\int_A^B F \cdot dR = \int_A^B F \cdot T \, ds$, where T is the unit tangent vector and s is the arc-length parameter for the path joining the points A and B. Thus the line integral representing work is also known as the integral of the tangential component of F along a path from A to B.

3. Show that $\displaystyle\int_A^B F \cdot dR = \int_A^B d\phi = \phi(B) - \phi(A)$ when F is conservative.

(Hint: use $F = \nabla\phi$ and differentials as discussed in Section 3.1)

4. Use the result of Exercise 3. to show that if F is conservative then the line integral of F around every closed path is zero.

5. Use the result of Exercise 4. to show that if F is conservative then

$$\int_A^B F \cdot dR$$

is independent of the path joining A and B.

6. Show that each of the following vector fields is conservative and evaluate

$$\int_A^B F \cdot dR$$

for the given points using the results of Exercise 3.

a) $F = yzi + xzj + xyk$ from $(1,1,3)$ to $(2,3,1)$.

b) $F = (3y + x^2)i + (5y^4 + 3x)j + (z^2 + 4\cos z)k$ from $(1,1,0)$ to $(2,3,\pi/2)$.

7. Use Green's Theorem to verify each of the following for any closed curve   C
    in the xy plane enclosing an area A.   The integration is in the counterclockwise
    sense in each case.

    a) $\int_C y \ dx = -A$ b) $\int_C x \ dy = A$

    c) $\int_C x \ dx = 0$ d) $\frac{1}{2} \int_C xdx - ydy = A$

8. Evaluate $\int_C F \cdot dR$ for each of the following when the direction on   C   is
    counterclockwise.

    a) $F = -4y\mathbf{i} + 3x\mathbf{j}$ and   C   is the circle $x^2 + y^2 = 10$.

    b) $F = y^2\mathbf{i} + (x^2 + xy)\mathbf{j}$ and   C   is the curve enclosing the region bounded
        by $y = x$ and $x = y^2$.

    c) $F = yx^2\mathbf{i} + y^2\mathbf{j}$ and   C   is the circle $x^2 + y^2 = 2$.   (Hint: use polar
        coordinates as in Section 4.2).

9. Suppose that $\nabla \times F$ is zero every where except at a point   P   inside the closed

    path C_1.   Show that $\int_{C_1} F \cdot dR = \int_{C_2} F \cdot dR$ for any other closed path also

    enclosing the point P.   The direction of integration on C_2 is in the same sense
    as on C_1.   (Hint: Consider Figure 2. and apply Green's theorem to the curve
    C   composed of C_1, ℓ, C_2 and ℓ again, with the opposite direction).

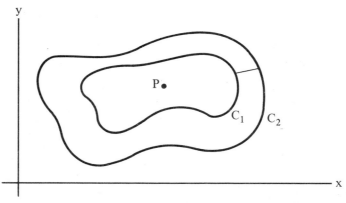

Figure 2.

10. Use the results of Exercise 9. to evaluate $\int_C F \cdot dR$ for each of the

following when C is taken in the counterclockwise direction.

a) $F = \dfrac{yi - xj}{x^2 + y^2}$ and C is the ellipse $x^2 + 2y^2 = 2$.

b) $F = \dfrac{xi + yj}{\sqrt{x^2 + y^2}}$ and C is the boundary of the triangle with vertices at

(-1,-2), (1,-1) and (2,5).

5.8   SURFACE INTEGRALS

The integral theorems of vector calculus, to be introduced in the next section, relate one and three dimensional integrals to <u>surface integrals</u>. Surface integrals are a generalization of double integrals that were introduced in Chapter 4.

One form of a surface integral that is useful in applications is obtained by considering the heat flowing across a given surface in three dimensional space. Referring to Figure 1, Q(x,y,z) represents, at each point in space, the direction

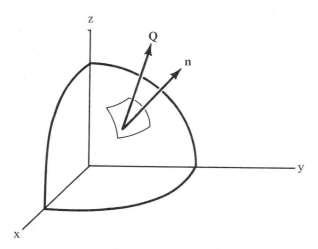

Figure 1.

and magnitude of heat flow (calories per unit area per unit time), and n is the unit vector perpendicular to the surface at each point. Thus the heat flowing across an element of area, denoted by ΔS_{ij}, is given approximately by

$$\Delta C_{ij} = Q(P_{ij}) \cdot n(P_{ij}) \, \Delta S_{ij} , \tag{1}$$

where P_{ij} is a point lying on ΔS_{ij}. If all such increments are added and the limit taken ($\Delta S_{ij} \rightarrow 0$ appropriately) then

$$C = \iint\limits_{S} Q \cdot n \, dS \tag{2}$$

gives the total heat flowing across the surface. The integral in Eq. (2) is known as a surface integral over the surface S. In order to evaluate it, the unit vector n must be known and the form of the area element dS must be known. Both of these will be discussed in detail in this section.

An alternate form of Eq. (2) is obtained by recalling that $Q \cdot n$ is a scalar and can be denoted as $g(x,y,z)$. Thus

$$C = \iint\limits_{S} g(x,y,z) dS \tag{3}$$

is also a surface integral of the function g over the surface S. It appears that the function to be evaluated in Eq. (2) or Eq. (3) ($Q \cdot n$ or $g(x,y,z)$) is a function of three variables in a two dimensional integral. However, it must be realized that $g(x,y,z)$ (or $Q \cdot n$) is evaluated only on a surface, which implies that one of the variables, say z, is a function of the other two. Thus g is then only a function of two variables. If $g(x,y,z) = 1$ in Eq. (3) we obtain

$$S = \iint\limits_{S} dS \tag{4}$$

as the surface area of S.

In working with surfaces in the xyz coordinate systems, the unit vector n can be found using techniques presented earlier in this chapter. If $G(x,y,z)$ equal to a constant specifies the surface, then

$$n = \frac{\nabla G}{|\nabla G|} \tag{5}$$

gives the desired vector, while if $z = f(x,y)$ specifies the surface then

$$n = \frac{\frac{\partial f}{\partial x} i + \frac{\partial f}{\partial y} j - k}{\sqrt{f_x^2 + f_y^2 + 1}} \quad . \tag{6}$$

There is one problem, however, in both Eqs. (5) and (6) concerning the correct direction. Each of the vectors n is perpendicular to the surface, but we could also say −n is perpendicular to the surface, thereby yielding a different value for the surface integral. Therefore, the appropriate direction must be specified. For a closed surface, such as a sphere, the outward or exterior vector is usually required.

The form of dS that appears in Eqs. (2), (3) and (4) can be found with the aid of Figure 2. Recall that dS represents the element of area on the surface. In the

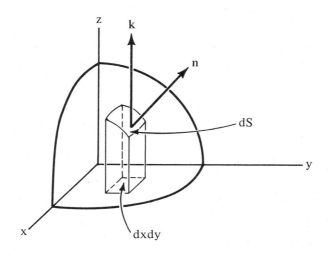

Figure 2.

xyz coordinate system this element can be projected onto one of the coordinate planes, usually the xy plane as shown in Figure 2. The projected area in the xy plane is given by dxdy and is related to dS by

$$dxdy \;=\; |\cos \gamma|\,dS, \tag{7}$$

where γ is the angle between the unit vectors n and k. If $\gamma = \pi/2$ then n is perpendicular to k and dS projects into a curve in the xy plane. In this case dS must be projected onto either the xz or yz planes, where the appropriate form of Eq. (7) becomes

$$dxdz \;=\; |\cos \beta|\,dS \tag{8}$$

or

$$dydz \;=\; |\cos \alpha|\,dS . \tag{9}$$

The angles α and β are the angles between n and i or j respectively.

Equation (7), (8) or (9) then yields the value for dS, depending on which coordinate plane is used in the calculations.

Example 1.   Evaluate $\iint\limits_{S} F \cdot n\ dS$ when $F = i + \frac{1}{2}xyj + zyk$,

S is the triangle with vertices (6,0,0), (0,3,0) and (0,0,2), and n is directed away from the origin on S.

The equation of the plane passing through the give points is
$x + 2y + 3z = 6$. Thus Eq. (5) is used to yield

$$n = \frac{i + 2j + 3k}{\sqrt{14}} \tag{10}$$

and from Section 5.2 we know that

$$\cos \gamma = n \cdot k = \frac{3}{\sqrt{14}} \tag{11}$$

since n and k are unit vectors. Thus

$$\iint\limits_{S} F \cdot n\ dS = \int_{0}^{3}\int_{0}^{6-2y} \frac{1}{\sqrt{14}}(1 + xy + 3zy) \frac{dxdy}{3/\sqrt{14}} . \tag{12}$$

where the limits of integration correspond to the region which is obtained when S is projected onto the xy plane, as shown in Figure 3.

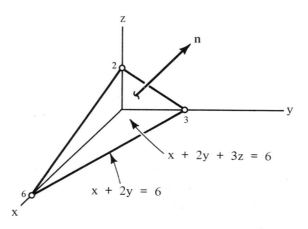

Figure 3.

On the surface S, $3z = 6 - x - 2y$ and thus Eq. (12) becomes

$$\iint\limits_{S} F \cdot n \, dS = \int_0^3 \int_0^{6-2y} \frac{1}{3}(1 + xy + y(6 - x - 2y))dxdy$$

$$= \frac{1}{3} \int_0^3 \int_0^{6-2y} (1 + 6y - 2y^2)dxdy, \tag{13}$$

where the constant $\sqrt{14}$ has been divided out since it appears in the numerator and denominator. The evaluation of Eq. (13) is accomplished using methods of Section 4.1, since it is an ordinary double integral, and yields

$$\iint\limits_{S} F \cdot n \, dS = \frac{1}{3} \int_0^3 (1 + 6y - 2y^2)(6 - 2y)dy$$

$$= \frac{2}{3} \int_0^3 (2y^3 - 12y^2 + 17y + 3)dy$$

$$= 12.$$

Example 2.   Evaluate $\iint\limits_{S} F \cdot n \, dS$ when $F = i + j + k$ and S is the open cylindrical surface shown in Figure 4, where the normal is indicated.

Since n does not vary continuously over the surface S (n has discontinuities at the corners) we must rewrite the surface integral as the sum of three integrals:

$$\iint\limits_{S} F \cdot n \, dS = \iint\limits_{S_1} F \cdot n_1 \, dS + \iint\limits_{S_2} F \cdot n_2 \, dS + \iint\limits_{S_3} F \cdot n_3 \, dS. \tag{14}$$

On the surface S_1 we have $n_1 = k$ and $dS = dxdy$ and thus

$$\iint\limits_{S_1} F \cdot n_1 \; dS \;\; = \;\; \iint\limits_{R} 1 \; dxdy \tag{15}$$

$$= \;\; \frac{\pi}{4} \; . \tag{16}$$

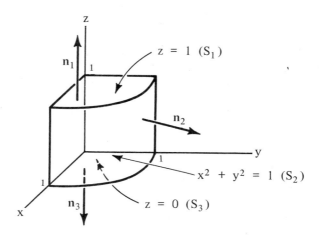

Figure 4.

The right hand integral of Eq. (15) is simply the area of the region R, the projection of S_1 onto the xy plane, and may be evaluated knowing the area of the quarter circle R. Likewise on the surface S_3 we have

$$n_3 = -k \quad \text{and} \quad dS = dxdy$$

so that

$$\iint\limits_{S_3} F \cdot n_3 \; dS \;\; = \;\; \iint\limits_{R} (-1) dxdy \;\; = \;\; -\frac{\pi}{4} \; . \tag{17}$$

Finally, for S_2 we see that the surface cannot be projected onto the xy plane since S_2 is everywhere perpendicular to the xy plane. Thus we obtain

$$n_2 \;\; = \;\; \frac{xi + yj}{\sqrt{x^2 + y^2}} \;\; = \;\; xi + yj$$

and

$$dS \;\; = \;\; \frac{dxdz}{|\cos \beta|} \;\; = \;\; \frac{dxdz}{|n_2 \cdot j|} \;\; = \;\; \frac{dxdz}{y}$$

since $x^2 + y^2 = 1$ on S_2. Hence

$$\iint\limits_{S_3} F \cdot n_3 \, dS = \int_0^1 \int_0^1 (x + y) \frac{dxdz}{y} \quad ,$$

where the limits are obtained from the projection of S_3 onto the xz plane. Solving for y in terms of x on the surface S_3 we obtain

$$\iint\limits_{S_3} F \cdot n_3 \, dS = \int_0^1 \int_0^1 (\frac{x}{\sqrt{1 - x^2}} + 1)dxdz \tag{18}$$

$$= \int_0^1 (- \sqrt{1 - x^2} + x)\Big|_0^1 \, dz = 2. \tag{19}$$

Substituting the values from Eqs. (16), (17) and (19) into Eq. (14) we arrive at

$$\iint\limits_{S} F \cdot n \, dS = \frac{\pi}{4} + 2 - \frac{\pi}{4} = 2$$

as the value of the desired integral.

The above examples have been worked in detail to illustrate the fine points that are involved when surface integrals are encountered.

In many problems the surfaces or functions F give rise to integrands which are more complicated than even that of Eq. (18), in which case a change of variables is necessary. An alternate approach is to write the surface S in parametric form, where x(u,v), y(u,v) and z(u,v) are written in terms of the two parameters u and v. In this case the position vector R:

$$R = x(u,v)\hat{i} + y(u,v)\hat{j} + z(u,v)k \tag{20}$$

describes the surface in three dimensional space when u and v are varied over the appropriate region of the uv space. If v is held constant, then R(u,v) is the position vector of a curve (as in Section 5.4) and thus $\frac{\partial R}{\partial u}$ is tangent to the surface since the curve v = constant must lie on the surface. Likewise, $\frac{\partial R}{\partial v}$ is tangent to the surface along the curve u = constant, as shown in Figure 5.

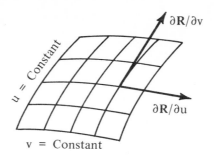

Figure 5.

The vector perpendicular to the surface may now be obtained by finding the cross product of the two tangent vectors. However, we also need an expression for the surface area dS, which may be obtained simultaneously with n by considering one differential element of area, as shown in Figure 6. In this case the area of

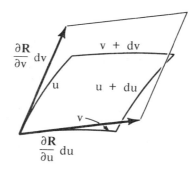

Figure 6.

the surface element is approximated by the area of the parallogram with sides parallel to $\frac{\partial R}{\partial u}$ du and $\frac{\partial R}{\partial v}$ dv respectively. In the limit this approximation becomes exact, and hence

$$dS = \left| \frac{\partial R}{\partial u} du \times \frac{\partial R}{\partial v} dv \right| \tag{21}$$

since the area of the parallelogram is the magnitude of the cross product (Section 5.3). Rewriting Eq. (21) in terms of vectors we obtain

$$n\ dS = \left(\frac{\partial R}{\partial u} \times \frac{\partial R}{\partial v} \right) dudv \tag{22}$$

since n is a unit vector in the direction of the cross product.

Example 3.   Evaluate $\iint\limits_{S} F \cdot n \, dS$ when $F = i + j + k$

> and S is the cylindrical portion S_2 of the surface in Figure 4.

For the cylindrical surface it is appropriate to use cylindrical coordinates. For the given surface then, $x = \cos \theta$, $y = \sin \theta$ and $z = z$, since $r = 1$ over the entire surface. Thus

$$R(\theta, z) = \cos(\theta)i + \sin(\theta)j + zk \tag{23}$$

is the parametric representation of the surface when $0 \leq \theta \leq \frac{\pi}{2}$ and $0 \leq z \leq 1$. The tangents are then obtained as

$$\frac{\partial R}{\partial \theta} = -\sin(\theta)i + \cos(\theta)j$$

and

$$\frac{\partial R}{\partial z} = k.$$

Thus

$$n \, dS = (\frac{\partial R}{\partial \theta} \times \frac{\partial R}{\partial z}) d\theta dz$$

$$= (\cos(\theta)i + \sin(\theta)j) d\theta dz,$$

where the correct direction for n may be verified. Hence

$$\iint\limits_{S} F \cdot n \, dS = \int_{0}^{1} \int_{0}^{\pi/2} (\cos \theta + \sin \theta) d\theta dz$$

$$= \int_{0}^{1} (\sin \theta - \cos \theta) \Big|_{0}^{\pi/2} dz = 2.$$

Notice that the same result was obtained in Example 2. In Example 3. the vector F was constant over the entire surface. If F were a function of x, y and z, then the new coordinates (in this case θ and z) must be substituted appropriately for x, y and z.

Exercises

1. Rework Example 1. by projecting the surface S onto the x plane.

2. Rework Example 2. by projecting the surface $x^2 + y^2 = 1$ onto the yz plane.

3. Evaluate $\iint\limits_{S}$ F · n dS   for each of the following cases.

    a) F = $x\hat{i}$ + $y\hat{j}$ + $z\hat{k}$, S is the sphere $x^2 + y^2 + z^2 = 2$ and n is directed
        outward.

    b) F = $2xz\hat{i} - 3yx^2\hat{j} + z^2\hat{k}$, S is the cube bounded by the coordinate planes
        and the planes x = y = z = 1, and n is directed outward.

    c) F = $z\hat{i} - y\hat{j} + x\hat{k}$, S is the triangle with vertices (1,0,0), (0,2,0),
        (0,0,3) and n is directed away from (0,0,0).

    d) F = $yz\hat{i}$ + k, S is the wedged shaped surface composed of z = y, z = 0,
        $y = 1 - x^2$ for y > 0 and n is directed outwards.

4. The surface z = f(x,y) can be written in parametric form as

$$R(x,y) \;=\; x\hat{i} + y\hat{j} + f(x,y)\hat{k} \;.$$

Use this form to find n as given in Eq. (6).

5. Find n dS for the general cylindrical surface $x^2 + y^2 = a^2$.

6. Find n dS for the general spherical surface $x^2 + y^2 + z^2 = a^2$.
    (Hint: refer to the spherical coordinates introduced in Section 4.2.)

7. Evaluate $\iint\limits_{S}$ F · n dS   when   F = $x\hat{i} - y\hat{j}$, S is the closed surface

bounded by z = 0, z = 2 and $x^2 + y^2 = 4$, and n is directed outward.

8. Evaluate $\iint\limits_{S}$ F · n dS   when   F = $\hat{i} + \hat{j} + \hat{k}$, S is the hemisphere

$z \;=\; \sqrt{9 - x^2 - y^2}$, and n is directed away from (0,0,0).

9.   Evaluate $\iint\limits_{S} F \cdot n \, dS$ when $F = xi + yj + zk$, S is the triangle with

vertices (1,0,0), (0,1,0), (0,0,1) and n is directed towards (0,0,0).

a)   Using rectangular coordinates.

b)   Using the parametric representation $x = u + v$, $y = u - v$, $z = 1 - 2u$.

5.9   THE DIVERGENCE THEOREM AND STOKES' THEOREM

This final section on vectors will introduce and discuss two results of fundamental importance in vector calculus. The first result is known as the Divergence
Theorem, which relates the surface integral of $F \cdot n$ over a closed surface, such
as a sphere or a rectangular box, to a volume integral over the entire volume enclosed by the surface. Mathematically, we require that F and $\nabla \cdot F$ be continuous,
in which case the Divergence Theorem yields

$$\iint\limits_{S} F \cdot n \, dS \;=\; \iiint\limits_{V} \nabla \cdot F \, dV. \tag{1}$$

Equation (1) assumes n is the outward normal to the surface S.

Example 1.   Evaluate $\iint\limits_{S} F \cdot n \, dS$ when $F = x^2 i + yzj + zk$

and   S   is the cube with sides $x = y = z = 0$ and
$x = y = z = 1$.

Since the cube is a closed surface and F is continuous we may apply
Eq. (1). To do this we calculate $\nabla \cdot F$, as given in Section 5.6, to
be

$$\nabla \cdot F \;=\; 2x + z + 1$$

which also is continuous. Thus

$$\iint\limits_{S} F \cdot n \, dS \;=\; \iiint\limits_{V} (2x + z + 1) dV$$

$$=\; \int_{0}^{1}\int_{0}^{1}\int_{0}^{1} (2x + z + 1) dz \, dy \, dx \;=\; \frac{5}{2} \;.$$

The derivation of Eq. (1) can be accomplished using the techniques of evaluating surface integrals developed in the last section. The surface integral in Eq. (1) can be written as

$$\iint\limits_{S} F \cdot n \, dS = \iint\limits_{S} F_1(n \cdot i) dS + \iint\limits_{S} F_2(n \cdot j) dS + \iint\limits_{S} F_3(n \cdot k) dS. \tag{2}$$

Now, for the third integral on the right hand side of Eq. (2) we assume S is composed of two portions S_1 and S_2 such that any line parallel to the z axis cuts each portion only once. If each surface S_1 and S_2 is divided into differential surface elements then there is a pair on each surface that is related as shown in Figure 1. These two elements contribute approximately

$$\Delta I = F_3(x,y,z_1)(n_1 \cdot k) dS_1 + F_3(x,y,z_2)(n_2 \cdot k) dS_2 \tag{3}$$

to the integral under discussion. In the previous section we saw that dS_1 and dS_2

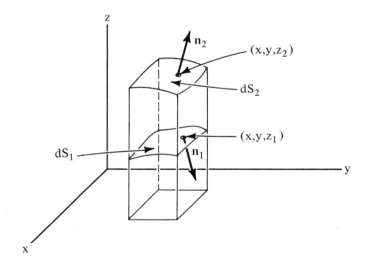

Figure 1.

are approximated by

$$dxdy = (n_2 \cdot k) dS_2 \tag{4}$$

and

$$-dxdy = (n_1 \cdot k) dS_1 . \tag{5}$$

As dS_1 and dS_2 approach zero the Eqs. (4) and (5) become exact. Thus ΔI in Eq. (3) becomes

$$\Delta I \ = \ F_3(x,y,z_1)(-dxdy) + F_3(x,y,z_2)(dxdy) \tag{6}$$

$$= \ (F_3(x,y,z_2) - F_3(x,y,z_1))dxdy. \tag{7}$$

If all the contributions ΔI are added and dS_1 and dS_2 allowed to approach zero, we then obtain

$$\iint\limits_{S} F_3(n \cdot k)dS \ = \ \iint\limits_{R} (F_3(x,y,z_2) - F_3(x,y,z_1))dxdy, \tag{8}$$

where R is the projection of S onto the xy plane. From the Fundamental Theorem of calculus we recognize that the integrand of Eq. (8) can be written as

$$F_3(x,y,z_2) - F_3(x,y,z_1) \ = \ \int\limits_{z_1(x,y)}^{z_2(x,y)} \frac{\partial F_3}{\partial z}(x,y,z)dz \tag{9}$$

and hence Eq. (8) yields

$$\iint\limits_{S} F_3(n \cdot k)dS \ = \ \iint\limits_{R} \int\limits_{z_1}^{z_2} \frac{\partial F_3}{\partial z}(x,y,z)dz \tag{10}$$

$$= \ \iiint\limits_{V} \frac{\partial F_3}{\partial z} dV. \tag{11}$$

Using similar arguments it can be shown that

$$\iint\limits_{S} F_1(n \cdot i)dS \ = \ \iiint\limits_{V} \frac{\partial F_1}{\partial x} dV \tag{12}$$

and that

$$\iint\limits_{S} F_2(n \cdot j)dS \ = \ \iiint\limits_{V} \frac{\partial F_2}{\partial y} dV. \tag{13}$$

Thus Eq. (2) yields

$$\iint\limits_{S} F \cdot n \ dS \ = \ \iiint\limits_{V} \left(\frac{\partial F_1}{\partial x} + \frac{\partial F_2}{\partial y} + \frac{\partial F_3}{\partial z}\right)dV, \tag{14}$$

which is the same as Eq. (1), the desired result.

We are now in a position to obtain a physical interpretation of the divergence that was introduced in Section 5.6. To obtain the desired interpretation consider a point (x,y,z) at which F and $\nabla \cdot F$ are continuous and consider a surface S enclosing this point. If the surface S is allowed to shrink to the given point, then we obtain from Eq. (1)

$$\lim_{V \to 0} \frac{\iiint\limits_{V} \nabla \cdot F \, dV}{V} = \lim_{V \to 0} \frac{\iint\limits_{S} F \cdot n \, dS}{V}, \tag{15}$$

where the division by the enclosed volume V is permissible before the limiting process is done. A close look at the left hand integral of Eq. (15) shows that

$$\lim_{V \to 0} \frac{\iiint\limits_{V} \nabla \cdot F \, dV}{V} = \nabla \cdot F \tag{16}$$

since as V approaches zero about the point (x,y,z), the integrand becomes a constant and thus we have

$$\lim_{V \to 0} \frac{\iiint\limits_{V} \nabla \cdot F \, dV}{V} = (\nabla \cdot F) \lim_{V \to 0} \frac{\iiint\limits_{V} dV}{V}. \tag{17}$$

Equation (17) then yields Eq. (16), since the numerator is equal to the denominator. Combining Eq. (16) and Eq. (15) we get

$$\nabla \cdot F = \lim_{V \to 0} \frac{\iint\limits_{S} F \cdot n \, dS}{V}, \tag{18}$$

which is the desired physical interpretation of the divergence. If F represents the velocity field for a fluid flow problem, then Eq. (18) shows us that the divergence of F at (x,y,z) represents the net flow of fluid out of an infinitesimal volume centered at (x,y,z). If the divergence is positive, then clearly the density at (x,y,z) is decreasing and conversely if the divergence is negative. Similar remarks can be made concerning the divergence of other vector fields.

The second result to be presented here is Stokes' Theorem, which relates a surface integral over an open surface, such as a hemisphere or the triangular section of a plane, to a line integral. As before, we require that F and $\nabla \times F$ be continuous, then Stokes' Theorem says that

$$\iint\limits_{S} \nabla \times F \cdot n \ dS \ = \ \int\limits_{C} F \cdot dR, \tag{19}$$

where C is the curve which forms the boundary of the open surface S and R is the position vector representing the curve C. The direction of integration for the line integral of Eq. (19) is determined by the orientation of the vector n. The direction of integration on C is such that if an observer walked along C with the area on the left his vertical position would be the same as the direction of the normal vector n.

Example 2.   Find the curve C and find the correct direction of integration for the application of Stokes' Theorem for the plane surface S shown in Figure 2.

The curve C, for use in Stokes' Theorem, is composed of the three straight line segments denoted as C_1, C_2 and C_3. The correct direction of integration is indicated by starting at (0,0,1) and proceeding to (3,0,0) along C_1, then along C_2 to (0,2,0) and finally back to (0,0,1) along C_3. Notice that the surface area of interest is always on the left as we proceed along C.

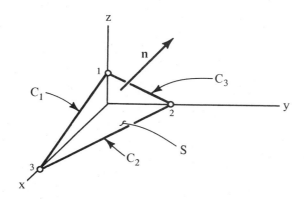

Figure 2.

It could happen, of course, that the normal to the surface S of Example 2. would be oriented in the opposite direction to that shown in Figure 2. The correct direction for the integration along C in that case would be just the opposite of that described in Example 2.

When S is a plane surface contained in the xy plane, then Eq. (19) reduces to Green's Theorem, which was discussed in Section 5.7. In this case, of course, n = k and C is the boundary curve in the xy plane with a counterclockwise direction of integration.

The actual use of Eq. (19) can be in either direction. That is, the integral $\int_C F \cdot dR$ may be evaluated as a surface integral using Eq. (19) or conversely the $\iint_S \nabla \times F \cdot n \, dS$ may be evaluated by reducing it to a line integral, again via Eq. (19). Both of these approaches are shown in the following example.

Example 3.

Evaluate $\iint_S \nabla \times F \cdot n \, dS$ when $F = (x+3y)i + (z-2x)j + (2z+1)k$, S is the hemisphere $z = \sqrt{1 - x^2 - y^2}$ and n is directed away from the origin.

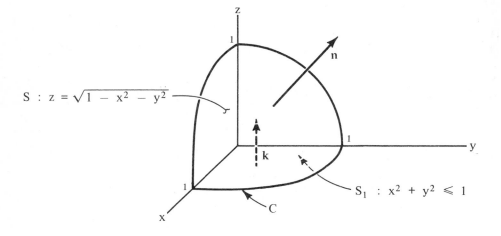

Figure 3.

The boundary curve C is the circle $x^2 + y^2 = 1$ in the xy plane (in the counterclockwise direction) and is shown in Figure 3. Hence,

$$\iint_S \nabla \times F \cdot n \, dS = \int_C F \cdot dR \qquad (20)$$

$$= \int_C (x + 3y)dx + (z - 2x)dy + (2z + 1)dz$$

$$= \int_C (x + 3y)dx - 2x\,dy \qquad\qquad (21)$$

since $z = 0$ and $dz = 0$ along C. This last integral can be evaluated
using the parametric representation $x = \cos \theta$, $y = \sin \theta$ for $0 \le \theta \le 2\pi$,
or by recognizing that the curve C is also the boundary curve for a
plane surface with normal $n = k$ (this new surface is the interior of the
circle $x^2 + y^2 = 1$). Thus Stokes' Theorem is used in reverse to yield

$$\int_C (x + 3y)dx - 2xdy = \int_C F \cdot dR$$

$$= \int\int\int_{S_1} \nabla \times F \cdot k\,dS \qquad\qquad (22)$$

since $n = k$ and S_1 is the region interior to $x^2 + y^2 = 1$ in the xy
plane. Notice that F in Eq. (22) is the same F as in Eq. (20)
evaluated on the new surface S_1. Calculating $\nabla \times F$ we obtain

$$\int\int_{S_1} \nabla \times F \cdot k\,dS = \int\int_{S_1} (-5k) \cdot k\,dS$$

$$= -5 \int\int_{S_1} dS = -10\pi \qquad\qquad (23)$$

since the area enclosed by S_1 is 2π. Using the final result of Eq. (23)
in Eqs. (22) and (20) we obtain

$$\int\int_S \nabla \times F \cdot n\,dS = -10\pi$$

as the value of our desired integral.

The relation of Stokes' Theorem and Green's Theorem is clearly shown in the last

example since the line integral of Eqs. (21) and (22) is precisely in the form
for use in Green's Theorem.

The derivation of Stokes' Theorem will not be given here since it really doesn't
lend itself to any geometrical or physical understanding of the final result (there
is a change of variables from x, y and z to u and v, then Green's Theorem and various
vector identities are used to get the final result). The physical meaning for the
curl of F is obtained in the same way that the physical meaning for the divergence
of F was obtained. Dividing both sides of Eq. (19) by S and letting S approach
zero about a point x,y,z we obtain

$$\lim_{S\to 0} \frac{\iint_S \nabla \times F \cdot n \, dS}{S} = \lim_{S\to 0} \frac{\int_C F \cdot dR}{S} \tag{24}$$

or

$$(\nabla \times F \cdot n) \lim_{S\to 0} \frac{\iint_S dS}{S} = \lim_{S\to 0} \frac{\int_C F \cdot dR}{S} \tag{25}$$

since as S → 0 $\nabla \times F \cdot n$ approaches a constant (at the point which S → 0) and can
be taken outside the integral. The limit on the left hand side is 1 and hence

$$\nabla \times F \cdot n = \lim_{S\to 0} \frac{\int_C F \cdot dR}{S} \tag{26}$$

which says that the <u>circulation</u> (the line integral of F around C) per unit area
is equal to the component of the curl in the direction of n. Notice that if a
different surface element were chosen (and hence a different R) a different limit
would be obtained on the right hand side of Eq. (26). From Eq. (26) we can deduce
that $|\nabla \times F|$, at a point, is equal to the maximum circulation about the point and
that the direction of $\nabla \times F$ is then the same as the normal to the surface element
that yields this maximum circulation.

Exercises

1. Derive Eq. (12).

2. Derive Eq. (13).

3. Evaluate $\iint\limits_S F \cdot n \, dS$ for each of the following.

    a) $F = 4xz\hat{i} + 2y\hat{j} - yz k$ and S is the surface of the box bounded by
        $x = y = z = 0$ and $x = 1$, $y = 2$, $z = 3$.

    b) $F = yz\hat{i} + xy\hat{j} + zk$ and S is the surface composed of the top $z = y + 2$
        and bottom $z = 0$ with sides $x = y = 0$ and $y = 1 - x^2$ in the first octant.

    c) $F = x\hat{i} + 2y\hat{j} - z k$ and S is the sphere $x^2 + y^2 + z^2 - 2z = 0$.

4. Show that $\iint\limits_S \nabla \times F \cdot n \, dS = 0$ whenever S is a closed surface.

5. Use Stokes' Theorem in evaluating $\iint\limits_S \nabla \times F \cdot n \, dS$ for each of the following.

    a) $F = xy\hat{i} - z^2\hat{j} + yk$, S is the triangular surface formed by $(1,0,0)$,
        $(0,2,0)$, $(0,0,1)$ and n is such that $n \cdot k > 0$.

    b) $\nabla \times F = (3x^2 - y^2)\hat{i} - (y + xz)\hat{j} + 2xyk$, S is the hemisphere

        $z = \sqrt{4 - x^2 - y^2}$ and n is such that $n \cdot k < 0$.

    c) $F = x\sin(y)\hat{i} - ze^x\hat{j} + xyzk$ and S is the ellipsoid $x^2 + 2y^2 + z^2 - 5x = 10$.

6. Evaluate $\int\limits_C F \cdot dR$ for each of the following.

    a) $F = z\hat{i} + (x + z)\hat{j} + (x + y)k$ and C is the curve of intersection of the
        cylinder $x^2 + y^2 = 2y$ and the plane $y = z$. Choose the direction of inte-
        gration on C to be counterclockwise when looking from the positive z
        direction.

    b) $F = 2xy\hat{i} + y\hat{j} + 5xzk$ and C is the curve of intersection of the cylinders
        $x^2 + y^2 = 4$ and $x^2 + z^2 = 4$ which passes through the first octant. Choose
        the direction on C which is clockwise when viewed from the positive z
        direction.

7. By letting $F = \nabla\phi$ in the Divergence Theorem show that

$$\iiint_V \nabla^2\phi \; dV = \iint_S \frac{\partial\phi}{\partial n} \; dS$$

where $\dfrac{\partial\phi}{\partial n}$, at any point on S, denotes the rate of change of ϕ in the direction of the outward normal to S at that point.

8. By letting $F = \Psi\nabla\phi$ in the Divergence Theorem show that

$$\iiint_V (\Psi\nabla^2\phi \; + \; \nabla\Psi \; \cdot \; \nabla\phi) dV \; = \; \iint_S \Psi \frac{\partial\phi}{\partial n} \; dS,$$

which is known as Green's First Identity.

9. Use the result of Exercise 8. to show that

$$\iiint_V (\Psi\nabla^2\phi \; - \; \phi\nabla^2\Psi) dV \; = \; \iint_S (\Psi \frac{\partial\phi}{\partial n} - \phi \frac{\partial\Psi}{\partial n}) dS,$$

which is Green's Second Identity.

10. If $F = \nabla \times G$, show that $\iint_S F \cdot n \; dS = 0$ for any closed surface S.

11. a) Evaluate $\iint_S F \cdot n \; dS$, as a surface integral, when $F = xy\mathbf{i} + yz\mathbf{j} + xz\mathbf{k}$

and S is the surface of a cube of side 2α centered at (x_0, y_0, z_0) and whose faces are parallel to the coordinate planes.

b) Divide the result of part a) by the volume of the cube and find the limit of the quotient as α approaches zero.

c) Calculate $\nabla \cdot F$ (F as in part a)) at (x_0, y_0, z_0) and compare to the result of part b).

12. a) Evaluate $\int_C F \cdot dR$, as a line integral, when $F = xy\hat{\imath} + yz\hat{\jmath} + xz\hat{k}$ and

C is the boundary of the plane square of side 2α whose center is located at (x_0, y_0, z_0) and whose normal is k.

b) Divide the result of part a) by the area of the square and find the limit of the quotient as α approaches zero.

c) Calculate $\nabla \times F \cdot n$ (F and n as in part a) at (x_0, y_0, z_0) and compare to the result of part b).

d) Repeat parts a), b) and c) when C is the boundary of the plane square of side 2α whose center is located at (x_0, y_0, z_0) and whose normal is j.

6

COMPLEX ANALYSIS

6.1 THE GEOMETRY OF COMPLEX NUMBERS

The subject of complex variables is one of the most important topics in mathematics for use in engineering and physics. The use of complex variables is standard in the solution of problems of heat flow, potential theory, electromagnetic theory, control and in many other areas. In this chapter we can only begin to discuss some of the concepts and results that are needed, but enough of an introduction will be given so that the concepts can be extended by the student as needed.

We have already encountered the idea of a complex number when we found roots of a quadratic equation involved in the solution of differential equations. To review, a complex number is of the form

$$z = a + bi \tag{1}$$

where a and b are real numbers and i is the imaginary number $\sqrt{-1}$. The real part of z (Re z) is a and the imaginary part of z (Im z) is b. Two complex numbers are equal if and only if the real and imaginary parts are equal.

Example 1.

$$2 + 3i = x + 3i \quad \text{only if} \quad x = 2$$

The real numbers we have used before are a subset of the complex numbers obtained by setting the imaginary parts equal to zero. Thus

$$z = x + 0i \tag{2}$$

for variable x is the set of all real numbers. The numbers

$$z = yi \tag{3}$$

are called <u>pure</u> <u>imaginary</u> numbers. The <u>complex conjugate</u> of z = a + bi is the complex number a - bi and is denoted by $\bar{z}$ (or z*).

In much of our work we will need the concept of a <u>complex variable</u>. In general
if z is a complex variable we will write

$$z = x + yi, \tag{4}$$

where x is the variable Re z and y is the variable Im z. This leads to the
graphical representation of complex numbers shown in Figure 1.

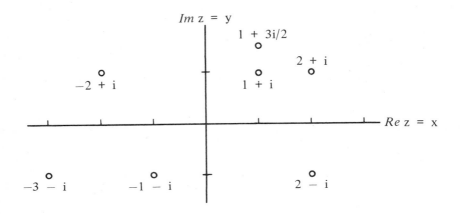

Figure 1.

Addition of complex numbers is accomplished by adding the real and the imaginary
parts respectively:

$$(a + bi) + (c + di) = (a + c) + (b + d)i. \tag{5}$$

Example 2.   The sum of z_1 = 2 + 3i and z_2 = -1 + 2i is given by

$$z_1 + z_2 = (2 + 3i) + (-1 + 2i)$$

$$= 1 + 5i.$$

The definition of addition for complex numbers just given is actually the same process as for the addition of vectors seen in the last chapter. This can be seen by
relating the Re z axis to the i direction for vectors and relating Im z axis to
the j direction for vectors. Equation (5) then becomes the same as vector addition.
Geometrically, using the points shown in Figure 1, it can be seen that the parallelogram law of vector addition also holds here.

Multiplication of a complex number by a real number s is done by multiplying
both the real and imaginary parts by s:

$$s(a + bi) = sa + sbi. \tag{6}$$

Example 3.   If $z = 2 - i$, then $3z$ is given by

$$3z = 3(2 - i)$$

$$= 6 - 3i.$$

The same relationship to vectors holds here, as essentially multiplication by a real number here is the same process as multiplication by a scalar for vectors.

Multiplication of two complex numbers is done by the following formula:

$$(a + bi)(c + di) = (ac - bd) + (bc + ad)i. \tag{7}$$

Example 4.   The product of $z_1 = 2 + i$ and $z_2 = 3 - i$ is given by

$$z_1 z_2 = (2 + i)(3 - i)$$

$$= (6 - (-1)) + (3 - 2)i$$

$$= 7 + i.$$

Division of complex numbers is possible, and the real and imaginary part of the quotient can be found by multiplying numerator and denominator by the conjugate of the denominator. This is illustrated in the following example.

Example 5.   If $z_1 = 2 - 3i$ and $z_2 = 3 + 4i$, then

$$z_1/z_2 \text{ is given by}$$

$$\frac{z_1}{z_2} = \frac{2 - 3i}{3 + 4i} = \frac{2 - 3i}{3 + 4i} \cdot \frac{3 - 4i}{3 - 4i}$$

$$= \frac{-6 - 17i}{9 + 16}$$

$$= -\frac{6}{25} - \frac{17}{25} i \; .$$

The geometric interpretation of multiplication and division will be given later after an alternate form of complex numbers has been developed. In summary, all the operations on complex numbers can be carried out by using the ordinary rules of algebra and the identity $i^2 = -1$. For example, in finding the product we have

$$(a + bi)(c + di) \; = \; ac + bci + adi + bdi^2$$

$$= \; (ac - bd) + (bc + ad)i,$$

which is exactly the same as Eq. (7).

The <u>absolute value</u> or <u>modulus</u> of a complex number a + bi is denoted as $|a + bi|$ and is given by

$$|a + bi| \; = \; \sqrt{a^2 + b^2} \; . \tag{8}$$

Example 6.   Find the absolute value of z = 2 + 3i.

Using Eq. (8) we have

$$|z| \; = \; |2 + 3i| \; = \; \sqrt{4 + 9} \; = \; \sqrt{13}.$$

The absolute value of z can be represented in terms of z and its conjugate as shown in the next example.

Example 7.   From the calculations

$$(a + bi)(a - bi) \; = \; a^2 + b^2 \; = \; |a + bi|^2$$

we see that if z = a + bi,

$$z\bar{z} \; = \; |z|^2. \tag{9}$$

The absolute value of a complex number represents the distance from the origin to the point representing the complex number in the complex plane of Figure 1. For this reason the absolute value of a complex number is also known as the magnitude of the complex number.

The polar and exponential form of complex numbers offer an alternate formulation as well as being simpler to handle in many cases. The polar form of z = x + yi is obtained by the "change of variables" x = r cos θ and y = r sin θ to yield

$$z \; = \; x + iy \; = \; r(\cos \theta + i\sin \theta). \tag{10}$$

From Figure 2. it is clear that:

$$r \; = \; \sqrt{x^2 + y^2} \; = \; |z|$$

and

$$\tan \theta \; = \; \frac{y}{x} \; . \tag{11}$$

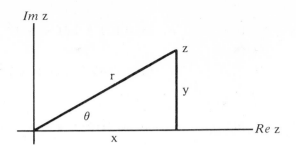

Figure 2.

The second equation in (11) is often rewritten

$$\theta = \tan^{-1}(y/x) = \arg z .\qquad(12)$$

The arg z is not a unique number, since by adding a multiple of 2π to arg z , we come back to the same point in the complex plane. This possibility must be watched very carefully when using complex variables. Also the quadrant in which $x + yi$ lies must be remembered, as the inverse tangent function is multiple-valued.

The polar form as given in Eq. (10) can be further changed using the identity

$$e^{i\theta} = \cos\theta + i\sin\theta,\qquad(13)$$

which we saw earlier in our work on differential equations (Section 1.2). Equation (13) cannot be justified until we consider functions of a complex variable; however, the usefulness of using Eq. (13) in Eq. (10) is so great, it is worth doing now. We obtain

$$z = x + iy = re^{i\theta},\qquad(14)$$

where r and θ are shown in Figure 2. and defined in Eqs. (11). Equation (14) gives the exponential form of a complex number. Its usefulness in calculating products and quotients can be seen by the following examples.

Example 8.   If $z_1 = r_1 e^{i\theta_1}$ and $z_2 = r_2 e^{i\theta_2}$ then

$$z_1 z_2 = r_1 e^{i\theta_1} \cdot r_2 e^{i\theta_2}$$

or

$$z_1 z_2 = r_1 r_2 e^{i(\theta_1 + \theta_2)}\qquad(15)$$

since exponents add when multiplying.

Example 8. verifies the important properties

$$|z_1 z_2| \; = \; |z_1||z_2| \tag{16}$$

and

$$\arg(z_1 z_2) \; = \; \arg z_1 \; + \; \arg z_2 \; . \tag{17}$$

Equation (15) also enables us to give a graphical interpretation of the product of two complex numbers, as shown in Figure 3.

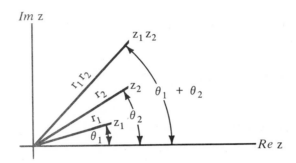

Figure 3.

Example 9.   If $z_1 \; = \; r_1 e^{i\theta_1}$ and $z_2 \; = \; r_2 e^{i\theta_2}$, then

$$\frac{z_1}{z_2} \; = \; \frac{r_1 e^{i\theta_1}}{r_2 e^{i\theta_2}} \; = \; \frac{r_1}{r_2} \, e^{i(\theta_1 - \theta_2)} \; .$$

As shown in Example 9, division using Eq. (14) is greatly simplified. Also we see that

$$\arg(z_1/z_2) \; = \; \arg z_1 \; - \; \arg z_2 \; , \tag{18}$$

which will be useful in our later work.

The exponential notation for complex numbers has a particular significance in electrical engineering problems, where the concept of a _phasor_ is frequently used. Very briefly, a phasor is denoted as $e^{-i\theta}$ and arises as a lag or delay in a sinusoidal time signal. Letting $\cos(\omega t - \theta)$ be the signal, we then have

$$\cos(\omega t - \theta) \; = \; \mathrm{Re}\{e^{i(\omega t - \theta)}\} \tag{19}$$

$$= \; \mathrm{Re}\{e^{i\omega t} e^{-i\theta}\}, \tag{20}$$

where Eq. (19) follows from Eq. (13). Thus Eq. (20) relates the phasor quantity, $e^{-i\theta}$, and the time signal $\cos(\omega t - \theta)$. Geometrically, the phasor represents an angle in the complex plane, such as was shown in Figure 3.

A final application of Eq. (14) is in finding the roots of complex or real numbers (for real numbers θ is either 0 or a multiple of π). If $z = re^{i\theta}$ then

$$z^{1/n} = (re^{i\theta})^{1/n}$$

$$= r^{1/n} e^{i\theta/n}$$

$$= r^{1/n} e^{(\theta + 2k\pi)i/n} , \quad k = 0,1\cdots n-1. \qquad (21)$$

Equation (21) follows since the arg z is only determined up to an additive multiple of 2π.

Example 10.   Find the three cube roots of -8.

Writing -8 in exponential form we get

$$-8 = 8e^{i\pi}$$

so that

$$(-8)^{1/3} = 2e^{(\pi + 2k\pi)i/3}, \quad k = 0,1,2. \qquad (22)$$

Thus for $k = 0$ we have $z_1 = 2e^{i\pi/3}$, for $k = 1$ we have $z_2 = 2e^{i\pi}$, and for $k = 2$ we have $z_3 = 2e^{5\pi i/3}$.

If larger values of k are used in Eq. (22) (or Eq. (21)) the first three roots (or n roots) will simply be repeated. For instance if $k = 3$ is used in Eq. (22) we would have

$$(-8)^{1/3} = 2e^{7\pi i/3}$$

$$= 2e^{2\pi i + \pi i/3}$$

$$= 2e^{i\pi/3}$$

since $e^{2\pi i} = 1$.

Some familiar regions in the plane can be described concisely by complex variables. For instance if z_0 is fixed and z variable then $|z - z_0| < C$ represents all points inside the circle of radius C with center at z_0 since

$$|z - z_0| = \sqrt{(x-x_0)^2 + (y-y_0)^2} \, .$$

Likewise $C_1 < |z| < C_2$ represents all points between concentric circles (an annular region), and Re $z \geq$ C represents all points to the right of the vertical line $x = C$ (a half plane). The exercises will illustrate some more regions.

Exercises

1. Let $z_1 = 3 + 2i$, $z_2 = -7 - i$ and calculate the following.

   a) $z_1 + z_2$ b) $z_1 z_2$

   c) z_1/z_2 d) $|z_1|$, $|z_2|$

   e) $3z_2 - z_1$ f) $\overline{z_1}\,\overline{z_2}$ and compare with (b)

2. Reduce each of the following expressions to the form $a + bi$.

   a) $2 + 3i - i^3$ b) $\dfrac{2 + i}{3 - i}$

   c) $\dfrac{i}{2 + i} - \dfrac{i - 2}{i}$ d) $i^{19} + 2(i + 1)^2$

3. If $z^2 = (\overline{z})^2$ show that either z is real or pure imaginary.

4. Where does z lie

   a) if $|z - 1| = 2$?   b) if Re $z = -2$?

   c) if Im $z = 3$?   d) if $|z + 1| = |z - 3|$?

   e) if arg $z = \pi/3$?   f) if Re$(z - 2) = |z|$?

5. Find the polar and exponential form for each of the following z.

   a) $z = 1 - \sqrt{3}i$ b) $z = i - 1$

   c) $z = 2$ d) $z = -3 - 4i$

6. Use the exponential form to calculate each of the following.

a) $\dfrac{1 + i}{1 - \sqrt{3}i}$

b) $(\frac{1}{2} + \frac{\sqrt{3}}{2} i)^4$

c) $(1 + i)^{1/3}$

d) $(1 - i)(-1 - \sqrt{3}i)$

e) $\sqrt{i}$

f) $2^{1/4}$

7. If z is a complex number lying outside the unit circle ($x^2 + y^2 = 1$) in the first quadrant, where does $1/z$ lie?

8. Describe geometrically the regions in the z plane determined by each of the following inequalities.

a) $1 < |z - i| < 3$

b) $|z - 2| > 1$

c) $0 \le \arg z \le \pi/2$

d) $-\pi < \arg z < \pi$

e) $|z| \le 1$

f) $|z - 1| < |z + 2|$

6.2 FUNCTIONS OF A COMPLEX VARIABLE AND THEIR DERIVATIVES

In the last section we introduced the concept of a complex variable as representing a variable point in the plane. If for each value of z there corresponds a value of w (a second complex variable), then w is called a complex function (or simply function) of z, normally written $w = f(z)$.

Example 1. $w_1 = z^2 + 1, w_2 = \dfrac{1}{z}$, and $w_3 = |z|$

are all functions of z since for each z the value of w may be found.

In general we assume that $w = f(z)$ is single-valued: for each z there is only one w. If more than one value of w can correspond to each z, we call $f(z)$ multiple-valued.

Example 2.   The function $w = z^{\frac{1}{2}}$ is multiple-valued since as we saw in the last section

$$w = z^{\frac{1}{2}} = r^{\frac{1}{2}} e^{i\theta/2}$$

or

$$w = z^{\frac{1}{2}} = r^{\frac{1}{2}} e^{i(\theta + 2\pi)/2}.$$

A multiple-valued function can be considered as a collection of single-valued functions, where each member is known as a _branch_ of the function. One branch is usually denoted as the _principal branch_, and the corresponding value is the _principal value_. This designation is made according to the problem being treated, not by universal convention.

Example 3.   For the function $w = z^{\frac{1}{2}}$, where $z = re^{i\theta}$, we may
      designate

$$w_1 = r^{\frac{1}{2}} e^{i\theta/2}, \qquad -\pi < \theta \leq \pi$$

as the principal branch while

$$w_2 = r^{\frac{1}{2}} e^{i\theta/2}, \qquad \pi < \theta \leq 3\pi,$$

is the other branch. Note that each branch is single-valued.
An alternate branch description for $w = z^{\frac{1}{2}}$ is given by

$$w_3 = r^{\frac{1}{2}} e^{i\theta/2}, \qquad 0 \leq \theta < 2\pi$$

and

$$w_4 = r^{\frac{1}{2}} e^{i\theta/2}, \qquad 2\pi \leq \theta < 4\pi,$$

since again each branch is single-valued.

For the first part of Example 3. the negative axis ($\theta = 0$, π, $3\pi, \cdots$) is known as the _branch cut_, while for the second half the positive axis ($\theta = 0$, 2π, $4\pi, \cdots$) is the branch cut. A branch cut is a line (or curve) across which θ is not continuous.
      A complex function is often considered to be a transformation or a _mapping_.
If $w = f(z)$ is a complex function with $z = x + iy$ and $w = u + iv$, then

$$u + iv = f(x + iy). \tag{1}$$

Equating real and imaginary parts of Eq. (1) we get

$$u = u(x,y) \quad \text{and} \quad v = v(x,y), \tag{2}$$

which shows that for each point in the xy plane there is a point in the "uv plane".

Example 4.   For the function

$$w = z^2 + 1$$

we have

$$u + iv = (x + iy)^2 + 1 = x^2 - y^2 + 1 + i2xy$$

and thus

$$u = x^2 - y^2 + 1, \quad v = 2xy. \tag{3}$$

The transformation in Eqs. (2) can be used to determine more than just points in the uv plane. For instance, in Example 4. the line $y = 1$ corresponds to the curve $u = x^2$, $v = 2x$ in the uv plane. These last two equations are the parametric equations for $u = v^2/4$. Thus the line $y = 1$ is transformed into the parabola $u = v^2/4$ by the function $w = z^2 + 1$.

The concepts of limit, continuity and differentiation of complex functions will now be developed. In general these concepts are similar to those for real functions of one variable, but since a mapping is involved they will be conceptually more difficult. We say that ℓ is the <u>limit</u> of $f(z)$ as z approaches z_0 if for any $\varepsilon > 0$ we can find a δ such that

$$|f(z) - \ell| < \varepsilon \quad \text{whenever} \quad 0 < |z - z_0| < \delta. \tag{4}$$

When Eq. (4) holds we write:

$$\lim_{z \to z_0} f(z) = \ell \tag{5}$$

and say $f(z)$ approaches ℓ when z approaches z_0. Equation (4) implies that the limit must be the same no matter how z approaches z_0. This can be seen geometrically in Figure 1. For multiple-valued functions the limit will depend on the particular branch. Some examples will be given to illustrate the concepts of a limit.

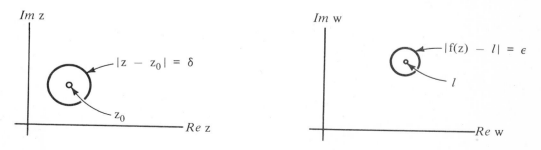

Figure 1.

Example 5.   The limit as z approaches i of the function
                    $z^3 + 1$ is given by

$$\lim_{z \to i} z^3 + 1 = 1 - i$$

                    since as z gets closer and closer to i, z^3
                    gets closer and closer to $i^3 = -i$.

Example 6.   For the function defined by

$$f(z) = \begin{cases} z^2 & z \neq 1 + i \\ 0 & z = 1 + i \end{cases}$$

    we have

$$\lim_{z \to 1+i} f(z) = (1 + i)^2. \tag{6}$$

Note that the limit given in Eq. (6) has nothing to do with the
functional value of f(z) at z = 1 + i, again since the limit of a
function only involves values of z close to the limit point
(1 + i in this case) and not the limit point itself.

Example 7.   The function $\dfrac{1}{z^2 - 1}$ does not have a limit at z = 1

                    since the denominator gets close to zero and hence
                    the quotient becomes undefined. In this case we
                    simply say that the limit does not exist.

If two functions f(z) and g(z) have limits, then so will their sum, product and
quotient (provided the denominator limit is not zero).

    The concept of continuity is very closely related to that of limit. If the
function f(z) has a limit at z = z_0 which is equal to the value of the function
at that point, then we say that f(z) is <u>continuous</u> at z_0. Mathematically we can
write

$$\lim_{z \to z_0} f(z) = f(z_0) \tag{7}$$

as the requirement to be fulfilled for a function to be continuous at a point z_0.
Note that the function must be defined at z_0 for continuity to hold.

Example 8.   The function $f(z) = z^2 + z + 1$ is continuous every-
                    where since Eq. (7) holds for all values of z.

Example 9.   The function $f(z) = \dfrac{z + 1}{z^2 + 1}$ is not continuous at

                    $z = \pm i$ since either the function is not defined or
                    the limit does not exist at $z = \pm i$. The function is
                    continuous everywhere else though.

Example 10.   The function defined by

$$f(z) = \begin{cases} \dfrac{z^2 - 1}{z + 1} & z \neq -1 \\[2mm] 0 & z = -1 \end{cases} \tag{8}$$

                    has a limit at $z = -1$, but the function is not con-
                    tinuous there since the value of the limit is not
                    equal to the value of the function at that point. The
                    function defined in Eq. (8) is continuous for all values
                    of z other than $z = -1$.

Example 10. is an example of a function that has a <u>removable</u> <u>discontinuity</u>. The
limit exists, but is not equal to the value of the function. We can "remove" the
discontinuity by redefining the function at $z = -1$ to be -2, the value of the limit.

    In the above discussions on limits and continuity we have assumed that the
points under discussion all had finite magnitudes and hence the concept of close-
ness as related to limits and continuity is appropriate. However, many times it is
necessary to work with the <u>point at infinity</u>. The point at infinity, denoted by ∞ ,
has the property that, for any z, $|z| < \infty$. That is, if the magnitude of z is
allowed to get larger, without bound, along any curve, then the point at infinity
is approached. This definition is motivated by the fact that if we let z approach
zero in the function $w = 1/z$, then the corresponding w will have arbitrarily large
magnitude. Thus to examine the limit or continuity of $f(z)$ at the point at infinity
(usually written $z = \infty$), set $z = 1/w$ and consider $f(1/w)$ for w near zero.

Example 11.   For the function

$$f(z) = \frac{z + 1}{z + 2}$$

    we have

$$f(1/w) = \frac{1/w + 1}{1/w + 2} = \frac{1 + w}{1 + 2w}$$

which is continuous at $w = 0$, so that $f(z)$ is continuous at $z = \infty$.

Although it is expedient to use "$z = \infty$", the point at infinity is not to be treated as a number, especially with algebraic expressions.

If two functions are continuous at a point z_0, then so are the sum, product and quotient, provided the denominator is not zero at the point. Thus polynomials are continuous and quotients of polynomials are continous, except where the denominator is zero.

The <u>derivative</u> of $w = f(z)$ is defined by:

$$\frac{df}{dz} = f'(z) = \lim_{\Delta z \to 0} \frac{f(z + \Delta z) - f(z)}{\Delta z} \tag{9}$$

provided $f(z)$ is single-valued in some region containing z. Since $f'(z)$ is defined using limits, Eq. (9) must hold no matter how Δz approaches zero. If Eq. (9) holds, we say $f(z)$ is differentiable.

Example 12.   For the function $f(z) = z^2$ we have

$$f'(z) = \lim_{\Delta z \to 0} \frac{(z + \Delta z)^2 - z^2}{\Delta z}$$

$$= \lim_{\Delta z \to 0} (2z + \Delta z) = 2z.$$

Example 13.   For the function $f(z) = |z|^2$ we have

$$f'(z) = \lim_{\Delta z \to 0} \frac{|z + \Delta z|^2 - |z|^2}{\Delta z}$$

$$= \lim_{\Delta z \to 0} \frac{(z + \Delta z)(\bar{z} + \overline{\Delta z}) - z\bar{z}}{\Delta z}$$

where the results of Example 7. of Section 6.1 have been used. Carrying out the multiplication and simplifying we get

$$f'(z) = \lim_{\Delta z \to 0} \bar{z} + \overline{\Delta z} + z \frac{\overline{\Delta z}}{\Delta z}$$

$$= \lim_{\Delta z \to 0} \bar{z} + \overline{\Delta z} + \bar{z} \frac{|\overline{\Delta z}| e^{-i\theta}}{|\Delta z| e^{i\theta}} \tag{10}$$

using the exponential form for Δz. Since $|\overline{\Delta z}| = |\Delta z|$ the limit in
Eq. (10) yields

$$f'(z) = \overline{z} + ze^{-2i\theta} \tag{11}$$

which says that the value of limit depends on how Δz approaches zero
since the approach direction determines θ. Hence the derivative does
not exist except at $z = 0$.

Example 12. illustrates that the formula for the derivative of a complex polynomial
is just like that for functions of real variables. In fact for all functions of a
complex variable we will use, the derivatives are obtained just as if z were real
instead of complex.

The following properties of the derivative can be verified using Eq. (9):

1) $\dfrac{d}{dz} (z^n) = nz^{n-1}$

2) $\dfrac{d}{dz} [f(z) + g(z)] = f'(z) + g'(z)$

3) $\dfrac{d}{dz} f(z) g(z) = f(z)g'(z) + f'(z)g(z)$

4) $\dfrac{d}{dz} \dfrac{f(z)}{g(z)} = \dfrac{g(z) f'(z) - f(z) g'(z)}{g^2(z)}$, $(g(z) \neq 0)$

5) $\dfrac{d}{dz} f[g(z)] = \dfrac{df}{dw} \dfrac{dg}{dz}$, $w = g(z)$.

We will discuss the concept of a derivative further in the next section by
finding the Cauchy-Riemann equations. These are easily used to verify that $f'(z)$
exists and they also play a very important role in many applications.

Exercises

1. Let $w = z^2 - 2$.

    a) Find w for $z = 1 + i$; $-2 - i$.

    b) Find all points in the w plane ($w = u + iv$) which correspond to the points
        which lie on $y = x$ in the z plane ($z = x + iy$).

c)   Repeat part b) for the curve $y = \dfrac{1}{x}$.

d)   Draw graphs in the   z   and   w   planes for parts b) and c).

2.   For each of the following functions find the real and imaginary parts as functions of x and y.

a) $w = z^2 + z - 1$ b) $w = \dfrac{1}{z}$

c) $w = \dfrac{1}{z} + z$ d) $w = \dfrac{z}{2 + z}$

3.   For each of the following functions, with the given regions, find the region in the   w   plane in which the corresponding values of w = f(z) lie.   Show the two regions graphically in each case.

a) $f(z) = 3z, \quad |\arg z| \leq \pi/3$

b) $f(z) = z^2, \quad |z| \geq 2$

c) $f(z) = 1/z, \; 1 < |z| < 2$

d) $f(z) = z^2 + 1, \quad |z| \leq 2$

e) $f(z) = \dfrac{z}{1 + z}, \quad |z| \leq 1$

4.   If $w = f(z) = (z^2 - 1)^{\frac{1}{2}} = (z + 1)^{\frac{1}{2}}(z - 1)^{\frac{1}{2}}$, then 1 and −1 are known as branch points of f(z).

a)   Write (z + 1) and (z − 1) in expontential form and then write f(z) in terms of the magnitudes and arguments of (z + 1) and (z − 1).

b)   Show that a complete circuit around both 1 and −1 produces no change in f(z), while a complete circuit around only one of them produces a change in f(z).   (A complete circuit is a closed path returning to the starting point.)

5. Evaluate each of the following limits, if they exist.

a) $\lim\limits_{z \to 2i} z^3 - 8z + 1$

b) $\lim\limits_{z \to i} \dfrac{z^4 + 2z^2 + 1}{z^2 + 1}$

c) $\lim\limits_{z \to 1+i} \dfrac{z^2 + 1}{z^2 - 2z + 2}$

d) $\lim\limits_{z \to i} f(z)$

where $f(z) = \begin{cases} 2z^2 + 4 & z \neq i \\ 0 & z = i \end{cases}$

e) $\lim\limits_{z \to \infty} \dfrac{z^2 + 2z + 1}{3z^2 + 4}$

f) $\lim\limits_{z \to 2i} \dfrac{2 - iz}{z^2 + 4}$

g) $\lim\limits_{z \to 3} \dfrac{z + 2}{z^2 - z - 6}$

h) $\lim\limits_{z \to \infty} \dfrac{3z - 6}{5z^2 - z + 3}$

6. a) For what value(s) of z, if any, is the function

$$f(z) = \dfrac{z^2 + 2z + 1}{z^2 + 4} \quad \text{not continuous?}$$

b) Same as part a) for $f(z) = \dfrac{z^2 + 5z + 4}{z^2 - 1}$

c) Do either of the functions in part a) or part b) have any removable discontinuities? If so, define a new function which agrees with the given function, except at the original discontinuity, and is continuous there.

7. Find $\dfrac{dw}{dz}$ for each of the following functions.

a) $w = 3z^2 + 2z + 1$

b) $w = (1 + 4z^2)^{\frac{1}{2}}$

c) $w = (z^2 - 1)^2 (z^3 - z + 1)^3$

d) $w = \dfrac{z + 1}{z - 1}$

8. Show that $f(z) = \bar{z}$ does not have a derivative anywhere.

6.3   THE CAUCHY–RIEMANN EQUATIONS AND ANALYTIC FUNCTIONS

In this section we shall discuss further some consequences of the definition of the derivative of a complex function that was introduced in the last section. The definition given previously:

$$\frac{df}{dz} = f'(z) = \lim_{\Delta z \to 0} \frac{f(z + \Delta z) - f(z)}{\Delta z} , \tag{1}$$

says that $f'(z)$ is the same no matter how Δz approaches zero, since it is given as a limit. The <u>Cauchy–Riemann</u> equations are obtained from Eq. (1) by letting $\Delta z \to 0$ different ways as shown in Figure 1. Thus if we let $\Delta z \to 0$ by first letting $\Delta y \to 0$ and

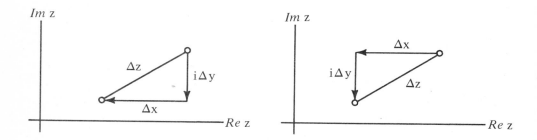

Figure 1.

then $\Delta x \to 0$ we obtain

$$f'(z) = \lim_{\Delta x \to 0} \frac{f(x + iy + \Delta x) - f(x + iy)}{\Delta x} \tag{2}$$

$$= \frac{\partial u}{\partial x} + i \frac{\partial v}{\partial x} \tag{3}$$

since $f(x + iy) = u(x,y) + iv(x,y)$. Similarly, if $\Delta x \to 0$ first and then $\Delta y \to 0$ we obtain

$$f'(z) = \lim_{i\Delta y \to 0} \frac{f(x + iy + i\Delta y) - f(x + iy)}{i\Delta y} \tag{4}$$

$$= -i \frac{\partial u}{\partial y} + \frac{\partial v}{\partial y} . \tag{5}$$

Since $f'(z)$ in Eq. (3) and Eq. (5) must be the same, if the derivative exists, we must conclude that

$$\frac{\partial u}{\partial x} = \frac{\partial v}{\partial y} \quad \text{and} \quad \frac{\partial v}{\partial x} = -\frac{\partial u}{\partial y} \tag{6}$$

by equating the real and imaginary parts. Conversely it can be shown that if $u(x,y)$ and $v(x,y)$ are single-valued, and if Eqs. (6) hold then $f'(z)$ must exist. Equations (6) are called the Cauchy-Riemann Equations.

Example 1.   For the function $f(z) = z^2$ we have

$$u = x^2 - y^2 \text{ and } v = 2xy.$$

Thus conditions (6) are satisfied since

$$\frac{\partial u}{\partial x} = 2x = \frac{\partial v}{\partial y}$$

and

$$\frac{\partial v}{\partial x} = 2y = -\frac{\partial u}{\partial y}.$$

Example 2.   For the function $f(z) = |z|^2$ we have

$$u = x^2 + y^2 \text{ and } v = 0.$$

Thus conditions (6) are not satisfied except at $z = 0$.

The Cauchy-Riemann equations also arise in a very natural way in a variety of applications. Using the notation of vector analysis developed in Chapter 5, we let $V = u(x,y)i - v(x,y)j$ be the velocity field for a two dimensional fluid motion. The motion is said to be incompressible if the divergence of V is zero and to be irrotational (no circulation) of the curl of V is zero. Since

$$\nabla \cdot V = \frac{\partial u}{\partial x} - \frac{\partial v}{\partial y}$$

and

$$\nabla \times V = -\left(\frac{\partial u}{\partial y} + \frac{\partial v}{\partial x}\right)k$$

we see that if the components of V satisfy the Cauchy-Riemann equations, then the fluid motion is incompressible (the reader is referred to the divergence theorem of Section 5.9) and irrotational (the reader is referred to Green's Theorem). Although the terminology is different, the same mathematical concepts could be developed if V were to represent an electric field for a two dimensional electromagnetic problem. Another very important application of the Cauchy-Riemann equations is to harmonic functions. This will be developed in the exercises.

A single-valued function $f(z)$ is analytic at a point z_0 if its derivative exists at every point in some neighborhood of z_0. (A neighborhood of z_0 is the set of points z satisfying $|z - z_0| < \delta$ for some $\delta > 0$). If $f(z)$ is analytic for every point except z_0 in a neighborhood of z_0, then z_0 is called an isolated singular point or an isolated singularity of $f(z)$.

Example 3.   The function

$$f(z) = \frac{1}{z^2 + 1}$$

is analytic except at $z = \pm i$; hence $z = \pm i$ are isolated singular points for $f(z)$.

Functions of a complex variable can also have non-isolated singularities. In fact, any point where $f(z)$ fails to be analytic is called a singularity. Example 4. illustrates a non-isolated singularity.

Example 4.   The derivative of the function

$$f(z) = z^{3/2}$$

$$= r^{3/2} e^{i3\theta/2} \qquad -\pi < \theta \leq \pi \qquad (7)$$

is given by

$$f'(z) = \frac{3}{2} z^{1/2}$$

$$= \frac{3}{2} r^{1/2} e^{i\theta/2}, \qquad -\pi < \theta < \pi. \qquad (8)$$

Thus for points along the negative axis and $z = 0$ the function given in Eq. (7) cannot be analytic since Eq. (8) will give different values when the points are approached from above ($\theta \to \pi$) and below ($\theta \to -\pi$). Nevertheless, the function $f(z)$ of Eq. (7) is analytic for all other values of z.

As with functions of real variables, if the derivative of $f(z)$ exists at a point z_0, then $f(z)$ is continuous at z_0. However, unlike functions of real-variables, it can be shown that if $f(z)$ is analytic at z_0, then all derivatives of $f(z)$ exist at z_0. This property has very important consequences, especially in the work on Taylor Series, which we will discuss in the next section.

In the remainder of this section we will discuss very briefly some frequently encountered analytic functions. Every polynomial in z, of the form

$$P(z) = a_n z^n + a_{n-1} z^{n-1} + \cdots + a_1 z + a_0, \tag{9}$$

where a_0, a_1, $\cdots$, a_n are complex constants, is analytic for all z. Every rational function in z, of the form

$$R(z) = \frac{a_n z^n + \cdots + a_1 z + a_0}{b_m z^m + \cdots + b_1 z + b_0}, \tag{10}$$

is analytic for all z except at the finite number of points where the denominator polynomial is zero. The fact that Eqs. (9) and (10) represent analytic functions can be verified with the aid of the properties of derivatives stated at the end of Section 6.2.

In many application problems, such as control system analysis and circuit analysis, the magnitude and argument of functions such as those represented by Eqs. (9) or (10) must be found. This can be accomplished with the use of the factored form of the polynomial

$$P(z) = a_n (z - z_1)(z - z_2) \cdots (z - z_n), \tag{11}$$

which exists by the Fundamental Theorem of Algebra. Each of the factors in Eq. (11) may now be written in the exponential form

$$z - z_k = r_k e^{i\theta_k}, \qquad k = 1, 2, \cdots n, \tag{12}$$

and hence

$$P(z) = a_n (r_1 e^{i\theta_1})(r_2 e^{i\theta_2}) \cdots (r_n e^{i\theta_n}). \tag{13}$$

Thus

$$|P(z)| = |a_n| \, r_1 r_2 \cdots r_n \tag{14}$$

and

$$\arg P(z) = \theta_1 + \theta_2 + \cdots + \theta_n, \tag{15}$$

since the exponents add when multiplying. A similar result for rational functions may also be found. One of the advantages of using Eqs. (14) and (15) is that a graphical approach can be used to calculate $|P(z)|$ and arg P(z), as shown in Example 5.

Example 5.   The function

$$R(z) = \frac{z + 1}{z^2 + 1}$$

can be written as

$$R(z) = \frac{r_3 e^{i\theta_3}}{r_1 e^{i\theta_1} \, r_2 e^{i\theta_2}} = \frac{r_3}{r_1 r_2} e^{i(\theta_3 - \theta_1 - \theta_2)}$$

where the r_n and θ_n for this problem are defined in Figure 2.

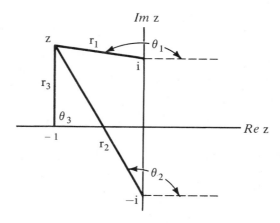

Figure 2.

The exponential function of a complex variable is defined by the following equation:

$$e^z = e^{x + iy} = e^x(\cos y + i \sin y). \tag{16}$$

Checking the Cauchy-Riemann equations with $u = e^x \cos y$ and $v = e^x \sin y$ we find that e^z is analytic for all finite values of z. Once we know that the derivative exists, we may use Eq. (3) to find the derivative of e^z:

$$\frac{d}{dz}(e^z) = \frac{\partial u}{\partial x} + i\frac{\partial v}{\partial x}$$

$$= e^x \cos y + i e^x \sin y.$$

Thus

$$\frac{d}{dz} (e^z) = e^z , \tag{17}$$

which is exactly the same result obtained for functions of a real variable. Using the definition (16) it is easy to show that e^z is never zero for any value of z and that e^z is periodic with period $2\pi i$. That is

$$e^{z+2\pi i} = e^z e^{2\pi i} = e^z . \tag{18}$$

The magnitude and argument of e^z are

$$|e^z| = e^x \tag{19}$$

and

$$\arg(e^z) = y, \tag{20}$$

which are also obtained using the definition. Other properties of the complex exponential function are exactly the same as for the real exponential function.

The inverse of the exponential function is the logarithm function

$$\ln z = \ln(re^{i\theta})$$
$$= \ln r + i\theta. \tag{21}$$

Since the arg $z = \theta$ is not uniquely defined, Eq. (21) implies that $\ln z$ is multiple valued, and in fact has infinitely many branches, each corresponding to a θ interval of length 2π. The $\underline{principal\ value}$ of $\ln z$ is usually given by

$$\ln z = \ln r + i\theta \qquad -\pi < \theta \leq \pi \tag{22}$$

$$= \frac{1}{2} \ln(x^2 + y^2) + i \tan^{-1}(y/x) \tag{23}$$

where $\tan^{-1}$ is between 0 and π if $y > 0$ and between $-\pi$ and 0 if $y < 0$. Using Eq. (23) it is easy to show that the Cauchy-Riemann equations are satisfied except at the origin and along the negative real axis (since the $\tan^{-1}(y/x)$ has a multiple limit for $y = 0$ and $x < 0$). Thus $\ln z$ is analytic for $-\pi < \theta < \pi$, $r > 0$. Since the Cauchy-Riemann equations are satisfied, we may find the derivative of $\ln z$ by using Eqs. (3) and (23)

$$\frac{d}{dz}(\ln z) = \frac{1}{2}\frac{\partial}{\partial x}\ln(x^2 + y^2) + i\frac{\partial}{\partial x}\tan^{-1}(y/x) \tag{24}$$

$$= \frac{x}{x^2 + y^2} - \frac{iy}{x^2 + y^2} = \frac{\bar{z}}{z\bar{z}}$$

$$= \frac{1}{z}, \tag{25}$$

which again is just like the derivative for the logarithm of a real variable.

The usual formulas for the logarithm of powers, quotients and products hold, provided one uses the appropriate branch. For instance

$$\ln z_1 + \ln z_2 = \ln r_1 + i\theta_1 + \ln r_2 + i\theta_2$$

$$= \ln r_1 r_2 + i(\theta_1 + \theta_2)$$

$$= \ln z_1 z_2, \tag{26}$$

provided the argument of $z_1 z_2$ is taken as $\theta_1 + \theta_2$. Using Eq. (16) and Eq. (21), the inverse function nature of $\ln z$ and e^z can be verified:

$$e^{\ln z} = z \quad \text{and} \quad \ln e^z = z. \tag{27}$$

Other functions of complex variables such as the trigonometric functions and hyperbolic functions can also be defined and handled similar to those discussed here.

As pointed out in the previous section, a function $w = f(z)$ can be interpreted as a transformation or mapping from the z plane to the w plane. If $f(z)$ is analytic and $f'(z) \neq 0$, such a mapping has an additional property: that of being conformal. A mapping is conformal if two curves in the z plane intersect at an angle ϕ and the corresponding curves in the w plane also intersect at the angle ϕ in the same sense.

Example 6.   The transformation $w = z + 2$ is conformal since it
                 represents a translation (all points in z plane are
                 shifted to the right two units) and hence all angles
                 will be preserved.

Conformal mapping is important in many applied problems since it is necessary to transform complicated regions into simpler ones.

We will only touch upon all the material that could be presented in the area of conformal mappings. The presentation here should be adequate for an understanding of the representation of the complex functions that we have introduced and should enable the student to use the elementary concepts involved. The general translation mapping is given by

$$w = z + \alpha, \tag{28}$$

where α is any complex number. In this case each point z is moved Re α units horizontally and Im α units vertically. The transformation

$$w = Az, \qquad A \neq 0, \tag{29}$$

for any complex A represents a rotation and stretching. If $z = re^{i\theta}$ and $w = \rho e^{i\phi}$ then $\rho = |A|r$ and $\phi = \arg(A) + \theta$ and thus all distances from the origin are "magnified" by $|A|$ and all points are rotated through an angle arg A. The equation

$$w = Az + \alpha, \qquad A \neq 0 \tag{30}$$

gives rise to a transformation which is equivalent to Eq. (29) followed by Eq. (28). The reciprocal transformation

$$w = \frac{1}{z}, \qquad z \neq 0 \tag{31}$$

involves both a reflection about the real axis and "inversion" in the circle of radius 1 about the origin, as shown in Figure 3. Regions outside the circle

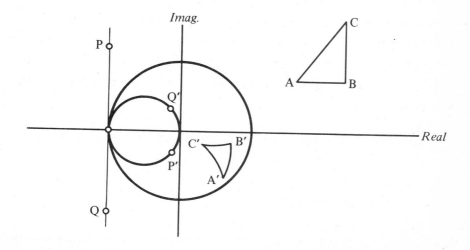

Figure 3.

correspond to smaller ones inside and vice-versa. It can be shown (Exercise 14.)
that circles (including straight lines as circles with infinite radius) transform
into circles. Notice in Figure 3. how the angles are preserved as the region ABC
is transformed into the region A'B'C' even though the straight lines of ABC do not
go into straight lines in A'B'C'.

The above four mappings are all special cases of the general linear fractional
transformation

$$w = \frac{az + b}{cz + d}, \qquad z \neq -d/c, \tag{32}$$

which is conformal provided $ad - bc \neq 0$ (w is a constant in this case). In addition,
the mapping accomplished by Eq. (32) can also be accomplished by a succession of
transformations of the previous types:

$$z_1 = cz + d, \qquad z_2 = \frac{1}{z_1}$$

and   (33)

$$w = \frac{a}{c} + \frac{bc - ad}{c} z_2 .$$

Since these all have the property of transforming circles into circles, then so does
the general transformation. One of the constants in Eq. (32) can be divided out,
leaving only three, and thus it is possible to find a linear fractional transformation which takes three given points z_1, z_2 and z_3 into three preassigned points
w_1, w_2 and w_3. Finally, it can be shown (Exercise 12.) that there are at most two
fixed points for Eq. (32) (except for the transformation $w = z$). A fixed point is
one which transforms into itself ($z = f(z)$).

The transformation $w = z^2$ is not conformal at $z = 0$, but is conformal everywhere
else. Thus angles will be preserved everywhere except at the origin, where they are
doubled (Exercise 13.).

The final conformal mapping to be presented is the exponential transformation

$$w = e^z . \tag{34}$$

If we set $w = \rho e^{i\phi}$ then we have that $\rho = e^x$ and $\phi = y$ and hence the lines given by
$x = $ constant are mapped onto circles and the lines given by $y = $ constant are mapped
onto rays (straight lines starting at the origin and going to ∞ at a constant
angle). A rectangular region $a \leq x \leq b$, $c \leq y \leq d$ is thus transformed into a
region bounded between two concentric circles and two rays as shown in Figure 4.
Every horizontal strip bounded by $y = c$ and $y = c + 2\pi$ in the z plane is transformed
onto the full w plane, since the arg $w = y$ completes a full cycle in this strip.

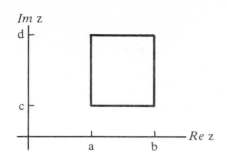 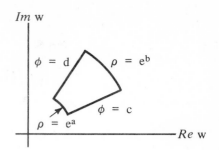

Figure 4.

Exercises

1. Use the Cauchy-Riemann equations to determine where each of the following functions is analytic.

a) $w = z^3 - 2z$ b) $w = \dfrac{1}{z}$

c) $w = \bar{z}$ d) $w = 2x + xy^2 i$

e) $w = \sin x \cos y + i \cos x \sin y$ f) $w = |x^2 - y^2| + 2i|xy|$

2. Let $w = f(z) = u(x,y) + i\, v(x,y)$, where u and v satisfy the Cauchy-Riemann equations. Show that $\dfrac{\partial^2 u}{\partial x^2} + \dfrac{\partial^2 u}{\partial y^2} = 0$ and that $\dfrac{\partial^2 v}{\partial x^2} + \dfrac{\partial^2 v}{\partial y^2} = 0$. Both u and v are called __harmonic functions__: continuous functions which are solutions of the partial differential equation $\dfrac{\partial^2 f}{\partial x^2} + \dfrac{\partial^2 f}{\partial y^2} = 0$ $(\nabla^2 f = 0)$.

Thus the real and imaginary parts of an analytic function are harmonic functions.

3. Show that $u = \dfrac{y}{x^2 + y^2}$ is a harmonic function (Exercise 2.) and use the Cauchy-Riemann equations to find $v(x,y)$ so that $u + iv$ is an analytic function. The functions u and v are then called conjugate harmonic functions.

4. Show that the Cauchy-Riemann equations become $\dfrac{\partial u}{\partial r} = \dfrac{1}{r}\dfrac{\partial u}{\partial \theta}$ and $\dfrac{1}{r}\dfrac{\partial u}{\partial \theta} = -\dfrac{\partial v}{\partial r}$ when u and v are expressed in polar coordinates r and θ. In this case

show that the derivative of $w = f(z)$ can be expressed as

$$\frac{dw}{dz} = e^{-i\theta} \left(\frac{\partial u}{\partial r} + i \frac{\partial v}{\partial r} \right) .$$

5. Use the results of Exercise 4. to show that each of the following functions is analytic.

a) $w = \ln r + i\theta$ b) $w = \dfrac{\cos \theta}{r} + i \dfrac{\sin \theta}{r}$

6. Let $f(z) = \dfrac{z + 2}{z(z^2 + 1)}$.

a) Find the magnitude and the argument of $f(z)$ at $z = 1 + i$, $-1 - i$, and $z = -1$.

b) Where is $f(z)$ not analytic?

7. Let $f(z) = \dfrac{1}{z^2 + 2z + 5}$ and consider $z = i\omega$, ω real.

a) Make a rough sketch of $|f(i\omega)|$ versus ω, as ω varies from zero to ∞.

b) Make a rough sketch of $\arg(f(i\omega))$ versus ω, as ω varies from zero to ∞.

8. a) Show that $e^{\bar{z}} = \overline{(e^z)}$.

b) Solve $e^z = -3$ for z in rectangular form, $z = x + iy$.

c) Evaluate $e^{-3 + i\pi/4}$.

d) Find $|e^{z^2}|$.

e) For what values of z, if any, is $|e^{-z}| < 1$?

9. a) Find the principal value of $\ln(1-i)$

b) Find all values of $\ln(i)$.

c) Solve $\ln z = i^{\pi/3}$, for z in rectangular form.

d) Show that all values of i^i are real numbers and find them.
        (Leave in terms of real exponentials.)

10. Show that the slope of the curve $u(x,y)$ = constant is given by

$$\frac{dy}{dx} = -\frac{\partial u/\partial x}{\partial u/\partial y},$$

(See Section 3.2). Thus show that, if u and v form a pair of conjugate harmonic functions (Exercise 3.), then the curves u = constant are perpendicular to the curves v = constant except at points where $f'(z) = 0$, $f(z) = u(x,y) + iv(x,y)$.

11. Show that the vector

$$V = (3 + \frac{x}{x^2 + y^2})i + (2 - \frac{y}{x^2 + y^2})j$$

represents the velocity field of an irrotational and incompressible fluid flow.

12. Show that Eq. (32) has at most two fixed points (except for the case w = z).

13. a) Find the curves in the w plane corresponding to x = 1 and y = 1 in the z plane under the transformation $w = z^2$ and show that the transformed curves intersect at right angles.

     b) Find the curves corresponding to y = x and x = 0 under the transformation $w = z^2$ and show that the angle of intersection in the w plane is twice that of the angle of intersection in the z plane.

14. a) Show that the equation of an arbitrary circle or straight line can be written in the form $az\bar{z} + bz + \bar{b}\bar{z} + c = 0$, a and c real.

     b) Use the result of part a) to show that circles or lines are transformed into circles or lines under the reciprocal transformation.

15. Show that the function $w = \frac{z}{1 - z}$ maps $|z| < 1$ onto the half-plane

     $\text{Re } w \geq -\frac{1}{2}$.

16. Find the linear fractional transformation that takes z = 0, ∞, 1 into w = ∞, 0, i respectively.

17. Find and graph the images of each of the following regions under the mapping $w = e^z$.

a) $-1 < x < 1,$ $-\pi/4 < y < \pi/4$

b) $0 < x < 1,$ $-\pi < y < \pi$

c) $-3 < x < -1,$ $0 < y < \pi/4$

6.4 POWER SERIES

The work with infinite series of complex numbers and complex variables closely follows that for real numbers and real variables and hence the student should review some previous work in this area. We will deal mainly with <u>power series</u>:

$$\sum_{n=0}^{\infty} a_n (z-c)^n = a_0 + a_1 (z-c) + a_2 (z-c)^2 + \cdots \tag{1}$$

where c is a fixed point in the complex plane and z is the complex variable.

Before we begin working with power series we must define what we mean by the convergence of an infinite series $c_1 + c_2 + c_3 + \cdots$ whose terms are complex numbers. If $c_n = a_n + ib_n$, where b_n and c_n are real numbers, the series $\sum_{n=1}^{\infty} c_n$ is said to converge if and only if the two real series $\sum a_n$ and $\sum b_n$ converge. If the last two series have sums A and B respectively (if the series converge they are said to sum or add to the number A or B) then

$$\sum_{n=1}^{\infty} c_n = C = A + iB. \tag{2}$$

The student will recall that convergence of a series can be defined in terms of a limit of the sequence of partial sums. For the complex series, if

$$C_n = c_1 + c_2 + \cdots + c_n, \tag{3}$$

then the infinite series $\sum_{n=1}^{\infty} c_n$ is said to converge to C if and only if $\lim_{n \to \infty} C_n = C.$

The series $\sum_{n=1}^{\infty} c_n$ is said to <u>converge absolutely</u> if the sum of absolute values

$|c_1| + |c_2| + |c_3| + \cdots$ converges. This is actually a stronger requirement than convergence since we are asking that an infinite sum of positive numbers add up to a finite value. However, absolute convergence is usually much easier to evaluate, as shown by the ratio test, which is reviewed here for reference.

Ratio test: For the series $\displaystyle\sum_{n=0}^{\infty} u_n$

$$\text{let } L \;=\; \lim_{n\to\infty} \left| \frac{u_{n+1}}{u_n} \right| \tag{4}$$

then

1) If $L < 1$, the series converges absolutely,

2) If $L > 1$, the series diverges,

3) If $L = 1$, or the limit does not exist, the test fails.

Returning now to the power series of Eq. (1), we could check convergence of this series by substituting in specific values of z and checking for convergence. However, a more pertinent approach is to apply the ratio test for arbitrary z and determine a region of z values for which the series converges. If we apply the ratio test to Eq. (1) with $u_n = a_n(z-c)^n$, we find that

$$L \;=\; \lim_{n\to\infty} \left| \frac{a_{n+1}(z-c)^{n+1}}{a_n(z-c)^n} \right|$$

$$=\; \lim_{n\to\infty} \left| \frac{a_{n+1}}{a_n} \right| \, |z-c| . \tag{5}$$

If we let

$$\rho \;=\; \lim_{n\to\infty} \left| \frac{a_n}{a_{n+1}} \right| \tag{6}$$

in Eq. (5), then $L = |z-c|/\rho$. Thus, the ratio test gives the important result that

1) If $|z-c| < \rho$ Eq. (1) converges absolutely, $\qquad\qquad$ (7)

2) If $|z-c| > \rho$ Eq. (1) diverges, $\qquad\qquad$ (8)

3) If $|z-c| = \rho$ no result, $\qquad\qquad$ (9)

where ρ is given in Eq. (6). The quantity ρ is called the radius of convergence, and Eq. (7) says that Eq. (1) will converge absolutely for all values of z within a circle of radius ρ about c, Eq. (8) says that Eq. (1) diverges for all points outside the circle and Eq. (9) says that for points on the circle, further investigations are necessary.

Example 1.   For the power series

$$\sum_{n=1}^{\infty} \frac{(z-1)^{n-1}}{n} = 1 + \frac{(z-1)}{2} + \frac{(z-1)^2}{3} + \cdots$$

ρ is given by

$$\rho = \lim_{n \to \infty} \left| \frac{1}{n} \Big/ \frac{1}{n+1} \right|$$

$$= \lim_{n \to \infty} \frac{n+1}{n} = 1.$$

Thus the series converges absolutely for $|z-1| < 1$ and diverges for $|z-1| > 1$. For points on the circle $|z-1| = 1$ the determination of convergence is much more tedious. For $z = 2$ we obtain the series $\sum_{n=1}^{\infty} 1/n$, which diverges, while for $z = 0$ we obtain the series $\sum_{n=1}^{\infty} (-1)^{n-1}/n$, which converges, but not absolutely.

For complex variables, a power series in negative powers often occurs

$$\sum_{n=1}^{\infty} \frac{b_n}{(z-c)^n} = \frac{b_1}{(z-c)} + \frac{b_2}{(z-c)^2} + \cdots \quad . \tag{10}$$

Applying the ratio test to these terms we find that Eq. (10) will converge if

$$|z-c| > \lim_{n \to \infty} \left| \frac{b_{n+1}}{b_n} \right| . \tag{11}$$

That is, a series of <u>negative</u> powers converges for all points <u>exterior</u> to a circle.
    There are many important properties of power series that we will not be able to discuss. Three that we will need are:

1)   A power series represents an analytic function at every point interior to its circle of convergence.

2)   A power series can be differentiated term by term at every point z interior to the circle of convergence.

3)   A power series may be integrated termwise at every point z interior to its circle of convergence.

Example 2.   Using the results of Example 1, we see that

$$f(z) = \sum_{n=1}^{\infty} \frac{(z-1)^n}{n} = (z-1) + \frac{(z-1)^2}{2} + \cdots$$

represents an analytic function for $|z-1| < 1$. Its derivative then is given by

$$f'(z) = \sum_{n=1}^{\infty} (z-1)^{n-1} = 1 + (z-1) + (z-1)^2 + \cdots .$$

The results in 1) and 2) above can be extended to series of negative powers. Care must be used in extending the third result to series of negative powers since the integral of $1/(z-c)$ introduces the $\ln(z-c)$ which has a branch cut starting at $z = c$.

The converse of the above results is to take a given function and expand it in a power series. If $f(z)$ is analytic in a circle of radius ρ about c, then from above it is appropriate to consider expanding $f(z)$ in a power series:

$$f(z) = \sum_{n=0}^{\infty} a_n (z-c)^n = a_0 + a_1 (z-c) + \cdots , \qquad (12)$$

which is assumed to converge for $|z-c| < \rho$. If $z = c$ in Eq. (12) we find $a_0 = f(c)$. If Eq. (12) is differentiated and $z = c$ we find $a_1 = f'(c)$. Continuing in this fashion we have:

$$f(z) = f(c) + f'(c) (z-c) + f''(c) \frac{(z-c)^2}{2} +$$
$$\qquad\qquad (13)$$
$$+ \cdots + f^{(n)}(c) \frac{(z-c)^n}{n!} + \cdots ,$$

which is the <u>Taylor Series for $f(z)$ about c</u>. Once Eq. (13) is found formally, as above, one can show that indeed it does converge in a circle of radius such that $f(z)$ is analytic. <u>No test for convergence is required for Eq. (13), however, since it can be shown that the radius of convergence is equal to the distance from c to the nearest singularity of $f(z)$</u>!

Example 3.   To find the Taylor series for $f(z) = e^z$ ($c = 0$) we recall that all derivatives of e^z are given by $f^{(n)}(z) = e^z$ and hence $f^{(n)}(0) = 1$ for all n. Thus

$$e^z = 1 + z + \frac{z^2}{2!} + \frac{z^3}{3!} + \cdots + \frac{z^n}{n!} + \cdots$$

or

$$e^z = \sum_{n=0}^{\infty} \frac{z^n}{n!} , \quad (0! = 1). \qquad (14)$$

This series converges for all z since e^z has no singularities for any finite z.

Example 4.   Find the Taylor series for $f(z) = 1/(1+z)$ about $c = 0, 1$.
The derivatives of $f(z)$ are given by

$$f'(z) = \frac{-1}{(1+z)^2}, \quad f''(z) = \frac{2}{(1+z)^3}, \quad \cdots \quad f^{(n)}(z) = \frac{(-1)^n n!}{(1+z)^{n+1}} \ .$$

Evaluating these at $z = 0$, we obtain

$$\frac{1}{1+z} = 1 - z + z^2 - z^3 + \cdots$$

$$= \sum_{n=0}^{\infty} (-1)^n z^n, \quad |z| < 1. \tag{15}$$

Evaluating the derivatives at $z = 1$ we obtain

$$\frac{1}{1+z} = \frac{1}{2} - \frac{1}{4}(z-1) + \frac{1}{8}(z-1)^2 + \cdots$$

$$= \frac{1}{2} \sum_{n=0}^{\infty} (-1)^n \frac{(z-1)^n}{2^n} \ , \quad |z-1| < 2. \tag{16}$$

The radius of convergence ($\rho = 1$) for Eq. (15) is the distance from $c = 0$ to the nearest singularity of $f(z)$, which is $z = -1$. The radius of convergence ($\rho = 2$) for Eq. (16) is obtained in a similar manner.

The series given in Eqs. (14) and (15) arise very frequently and should be learned by the student. The radius of convergence for the series of Eqs. (14), (15) and (16) can be verified using Eqs. (6) and (7).

Functions that are not analytic at a point c do not have Taylor series about c. However, if one considers negative powers, then we may try

$$f(z) = \sum_{n=0}^{\infty} a_n (z-c)^n + \sum_{n=1}^{\infty} \frac{b_n}{(z-c)^n} \ , \tag{17}$$

which is called a <u>Laurent</u> series and converges for an annular region $\rho_0 < |z-c| < \rho_1$ in which $f(z)$ is analytic. Notice that the Laurent series is composed of two sums, the one with negative powers converging for $|z-c| > \rho_0$ and the one with positive powers converging for $|z-c| < \rho_1$. If $\rho_0 < \rho_1$, then the series of Eq. (17) will converge for the annular region given, which is the area between the concentric circles $|z-c| = \rho_0$ and $|z-c| = \rho_1$. This region of course excludes the point $z = c$, at which $f(z)$ is assumed to be nonanalytic. We are assuming in Eq. (17) that $z = c$ is an isolated singularity of $f(z)$.

The coefficients a_n and b_n of Eq. (17) are given in terms of integrals, which we have not yet discussed. However, as the following examples show, for many problems we can find a_n and b_n using the results of Eq. (15).

Example 5.   Expand $f(z) = \dfrac{1 + z}{2z^2 + z^3}$ in powers of z $(c = 0)$.

In order to use Eq. (15) we must algebraically manipulate $f(z)$ to obtain a factor of the form of $1/(a+z)$. Thus

$$\frac{1 + z}{2z^2 + z^3} = \frac{1}{z^2}\left(\frac{1 + z}{2 + z}\right)$$

$$= \frac{1}{z^2}\left(1 - \frac{1}{2 + z}\right). \tag{18}$$

Now

$$\frac{1}{2 + z} = \frac{1}{2\left(1 + \frac{z}{2}\right)}$$

$$= \frac{1}{2}\sum_{n=0}^{\infty} (-1)^n \left(\frac{z}{2}\right)^n, \quad \left|\frac{z}{2}\right| < 1.$$

Substituting this into Eq. (18) we obtain

$$\frac{1 + z}{2z^2 + z^3} = \frac{1}{z^2}\left(1 - \frac{1}{2}\left(1 - \frac{z}{2} + \frac{z^2}{4} - \cdots\right)\right)$$

$$= \frac{1}{2z^2} + \frac{1}{4z} - \frac{1}{8} + \frac{z}{16} - \cdots$$

negative power of z

which converges for $0 < |z| < 2$.

Notice that in Example 5. there are only 2 terms with negative powers of z. Series with only a few negative powers occur frequently in practice. If

$$f(z) = \frac{b_n}{(z-c)^n} + \cdots + \frac{b_{-1}}{(z-c)} + a_0 + a_1(z-c) + \cdots \tag{19}$$

converges for $0 < |z-c| < R$, then $f(z)$ is said to have a <u>pole</u> of <u>order</u> n at $z = c$, provided $b_n \neq 0$. Thus the function of Example 5. has a pole of order 2 at $z = 0$.

If the number of non-vanishing coefficients in Eq. (19) becomes infinite, then $f(z)$ is said to have an <u>essential</u> singularity at $z = c$ (Eq. (19) still must converge for $0 < |z-c| < R$).

Example 6.   Using the results of Example 3. we obtain

$$e^{1/z} = 1 + \frac{1}{z} + \frac{1}{2!z^2} + \cdots + \frac{1}{n!z^n} + \cdots$$

$$= \sum_{n=0}^{\infty} \frac{z^{-n}}{n!} , \quad 0 < |z|.$$

Thus $e^{1/z}$ has an essential singularity at $z = 0$ (or e^z has an essential singularity at $z = \infty$).

The next example will illustrate a case where there are an infinite number of negative powers, but the function cannot be said to have an essential singularity.

Example 7.   Find a series for $f(z) = \dfrac{1}{z^2(z^2-1)}$ valid for

$|z| > 1$.

Again we wish to algebraically manipulate $f(z)$ so that appropriate use may be made of Eq. (15). Thus

$$\frac{1}{z^2(z^2-1)} = \frac{1}{z^4}\left(\frac{1}{1 - 1/z^2}\right).$$

Now, if $|z| > 1$ we know that $|1/z| < 1$ and thus Eq. (15) yields

$$\frac{1}{z^2(z^2-1)} = \frac{1}{z^4}\left(1 + \frac{1}{z^2} + \frac{1}{z^4} + \cdots\right)$$

$$= \frac{1}{z^4} + \frac{1}{z^6} + \frac{1}{z^8} + \cdots$$

$$= \sum_{n=0}^{\infty} \frac{1}{z^{2n+4}} , \quad |z| > 1. \tag{20}$$

The function $f(z)$ of Example 5. does not have an essential singularity at $z = 0$, however, since Eq. (20) does not converge for $0 < |z| < 1$. Exercise 5. will develop this point further.

This has been a very brief introduction to the concept of series representation of complex functions. However, with the aid of the worked examples, many problems encountered in practice can be handled. Also, many of the items may be familiar from previous work with functions of a real variable.

Exercises

1. Use Eq. (14) to show that $e^{iy} = \cos y + i \sin y$.
   (Hint: set $z = 0 + iy$ and review the power series for $\cos y$
          and $\sin y$ when y is a real variable).

2. Find the Taylor series for $\ln z$ about $c = 1$. What is its radius of convergence?

3. Find the Taylor series for $\ln(1 + z)$ about $c = 0$ by using Eq. (15) and integrating (which is permissible for $|z| < 1$). How do you evaluate the constant of integration?

4. a) Find a series for $\frac{1}{z+4}$ valid for $|z| < 4$.

   b) Find a series for $\frac{1}{z+4}$ valid for $|z| > 4$.

   c) Find a series for $\frac{z}{z+4}$ valid for $|z| < 4$, using part a).

5. Find a series for the function of Example 7. valid for $0 < |z| < 1$. Using this result determine whether the function has a pole or essential singularity at $z = 0$. If it has a pole determine its order.

6. For what values of z do each of the following series converge?

   a) $\sum_{n=1}^{\infty} \frac{(-1)^n (z-2)^n}{n 2^n}$

   b) $\sum_{n=0}^{\infty} n(z - i)^n$

   c) $\sum_{n=0}^{\infty} \frac{(z+3)^n}{(2n)!}$

   d) $\sum_{n=0}^{\infty} \frac{(2z-i)^n}{(n+1)^2}$

7. Show that $f(z) = \sum_{n=0}^{\infty} \frac{(2z)^n}{n!}$ satisfies $\frac{df}{dz} = 2f(z)$.

8. Show that $f(z) = \sum_{n=0}^{\infty} \frac{(-1)^n z^{2n}}{(2n)!}$ satisfies $\frac{d^2f}{dz^2} = -f(z)$.

9. Expand each of the following functions in a Laurent series which will converge for $0 < |z| < R$ and determine R. Also determine whether there is a pole (and its order) or an essential singularity at $z = 0$.

a) $\dfrac{1}{z^3(z-3)}$ b) $\dfrac{3}{2z^3 + z}$

c) $\dfrac{e^{-z}}{z}$ d) ze^{1/z^2}

10. Expand $\dfrac{z^2}{(1-z)(z+2)}$ in a series valid for each of the following regions.

a) $|z| < 1$ b) $1 < |z| < 2$ c) $|z| > 2$

(HINT: Use $\dfrac{z^2}{(1-z)(z+2)} = \dfrac{z^2}{3}\left(\dfrac{1}{z+2} - \dfrac{1}{z-1}\right)$ and write enough terms to

show the general trends.)

11. Expand $\dfrac{1}{z^3 - 2z^2 - 3z}$ in a series valid for each of the following regions.

a) $0 < |z| < 1$ b) $1 < |z| < 3$ c) $|z| > 3$

6.5 COMPLEX INTEGRATION

Some of the terminology and high points of complex integration will be discussed here. We will in no way be able to do justice to all the areas of integration of complex functions.

The definite integral of a complex function

$$I = \int_{\alpha}^{\beta} f(z)dz \tag{1}$$

looks very much like an integral of a real function. However, since α and β are complex numbers, there is an infinite choice of paths one may choose in going from α to β. Thus Eq. (1) is also called a <u>line</u> integral, which in general depends on the path chosen. Equation (1) can be rewritten in terms of real line integrals by writing $f(z)$ and dz in terms of their real and imaginary parts:

$$\int_{\alpha}^{\beta} f(z)dz = \int_{C} [u(x,y)dx - v(x,y)dy] + i \int_{C} [v(x,y)dx + u(x,y)dy], \qquad (2)$$

where C denotes the path joining α and β, with the direction of integration going from α to β. The direction must be specified if $\int_{C} f(z)dz$ is the notation used.

The two integrals on the right hand side of Eq. (2) are real line integrals and thus the material here closely parallels that of Section 5.7. In general they are evaluated by using real parametric equations

$$x = \phi(t) \quad \text{and} \quad y = \psi(t) \qquad (3)$$

for the curve C to reduce the line integrals to regular integrals of a function of one variable. For the first integral in Eq. (2) we would have

$$\int_{C} u(x,y)dx - v(x,y)dy = \int_{t_{\beta}}^{t_{\alpha}} [u(\phi(t),\psi(t))\phi'(t)dt - v(\phi(t),\psi(t))\psi'(t)dt] \qquad (4)$$

$$= \int_{t_{\beta}}^{t_{\alpha}} f(t)dt, \qquad (5)$$

when Eqs. (3) are used. In Eq. (5) t_{α} and t_{β} are the values of t such that Eqs. (3) yield α and β respectively and

$$f(t) = u(\phi(t),\psi(t))\phi'(t) - v(\phi(t),\psi(t))\psi'(t). \qquad (6)$$

A similar procedure is used for the second half of Eq. (2).

Example 1.   Evaluate $I = \displaystyle\int_{0}^{2+i} z^2 dz$ along the curve $y = x^2/4$,

as shown in Figure 1.

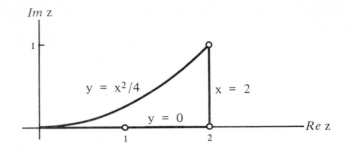

Figure 1.

Using the real and imaginary parts of z and dz, we find that

$$I = \int_{(0,0)}^{(2,1)} (x^2 - y^2)dx - 2xydy + i \int_{(0,0)}^{(2,1)} 2xydx + (x^2 - y^2)dy.$$

The parametric equations for the parabola are

$$x = t, \quad y = t^2/4$$

and thus

$$I = \int_0^2 (t^2 - t^4/16)dt - (2t)(t^2/4)(t/2)dt +$$

$$+ i \int_0^2 (2t)(t^2/4)dt + (t^2 - t^4/16)(t/2)dt$$

$$= \int_0^2 (t^2 - 5t^4/16)dt + i \int_0^2 (t^3 - t^5/32)dt$$

$$= 2/3 + i \; 11/3.$$

Example 2.   Evaluate the integral of Example 1. along the straight
                        line segments y = 0 and x = 2, also shown in Figure 1.
   From Example 1. we have that

$$J = \int_{(0,0)}^{(2,1)} (x^2 - y^2)dx - 2xydy + i \int_{(0,0)}^{(2,1)} 2xydx + (x^2 - y^2)dy.$$

Now along $y = 0$, $dy = 0$ and thus the integral along this part of
of the path is

$$J_1 = \int_0^2 x^2 \, dx + 0i,$$

where we have used x as the parameter. On $x = 2$, $dx = 0$ and
thus the integral on this part of the path is

$$J_2 = \int_0^1 - 4y \, dy + i \int_0^1 (4 - y^2) \, dy$$

where y is the appropriate parameter. Hence

$$J = J_1 + J_2 = 2/3 + 11i/3.$$

Example 2. illustrates the important result that _if the integrand of_ Eq. (1) _is
analytic in a region containing the path from_ α _to_ β, _then the value of the integral
is independent of the choice of the path_. As we shall see, this has important
consequences.

The following properties of complex integrals are valid and are similar to
properties of real integrals:

$$\int_\beta^\alpha f(z) \, dz = - \int_\alpha^\beta f(z) \, dz \quad \text{(along the same path)} \tag{7}$$

$$\int_\alpha^\beta k \, f(z) \, dz = k \int_\alpha^\beta f(z) \, dz \tag{8}$$

$$\int_\alpha^\beta [f(z) + g(z)] \, dz = \int_\alpha^\beta f(z) \, dz + \int_\alpha^\beta g(z) \, dz \tag{9}$$

$$\int_C f(z) \, dz \leq \int_C |f(z)| \, |dz| \leq M \int_C |dz| = ML \tag{10}$$

$$\int_{C_3} f(z) \, dz = \int_{C_1} f(z) \, dz + \int_{C_2} f(z) \, dz, \tag{11}$$

where C_3 is the two paths C_1 and C_2 joined, assuming C_1 and C_2 have only one point in common. It was this property we used in Example 2, where J_1 denoted the integral along C_1, the line y = 0, and J_2 denoted the integral along C_2, the line x = 2. The result of Eq. (7) can be rewritten as

$$\int_C f(z)dz = -\int_{-C} f(z)dz \tag{12}$$

where $-C$ is the notation for the same path as C but with the direction of integration reversed. In Eq. (10), $|f(z)| \leq M$ for all z along C and L is the length of the curve C.

Using property (7) and the result mentioned previously we see that

$$\int_C f(z)dz = 0, \tag{13}$$

when C is a <u>closed</u> curve and f(z) is single-valued and analytic inside and on C. This result is known as the Cauchy-Goursat Theorem. Unless otherwise noted, the direction of integration on a closed curve is in the counterclockwise sense. In evaluating integrals as in Example 1, Eq. (13) can be used to simplify the path of integration, as was done in Example 2. In addition, Eq. (13) has many important consequences in engineering practice and theory.

If f(z) has one or more singular points interior to the curve C, then Eq. (13) no longer holds in general. If the singular point arises because f(z) fails to be analytic at exactly one point, but is single-valued, and analytic everywhere else inside C, then the closed contour integration (the integral of f(z) around a closed path) can be handled by the use of <u>multiply connected</u> regions. A region R is <u>simply connected</u> if every closed curve within it encloses only points of R. Otherwise it is a multiply connected region.

Example 3.   The following regions illustrate various possibilities,
                      which are shown in Figure 2.

R_1: $|z| < 1$ is simply connected (a disk)

R_2: $0 < |z| < 1$ is multiply connected (a perforated disk)

R_3: $1 < |z| < 2$ is multiply connected (an annulus).

By adding an appropriate line joining the "inner" and "outer" curves of R_3, that region can be made simply connected as shown in Example 4. and Figure 3.

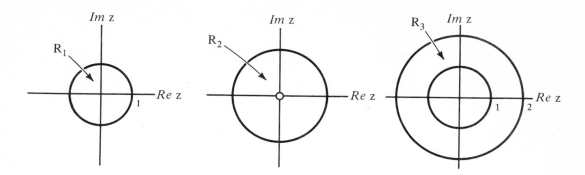

Figure 2.

Example 4. R_4: $1 < |z| < 2$ and $x \neq y$, $x > 0$

Thus every closed curve in R_4 encloses only points in R_4 since the curve is not allowed to cross the $y = x$ line for $x > 0$.

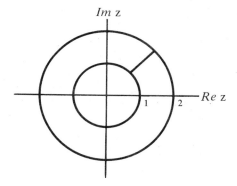

Figure 3.

Now, suppose it is necessary to evaluate $\int_C f(z)dz$ when $f(z)$ has an isolated singularity at $z = 0$ and C is the path $|z| = 2$. In this case Eq. (13) does not hold and the integral must be evaluated as a line integral, which can be fairly complicated for the given path C. However, with the aid of the region of Example 4, we can find a new path which will simplify the integration. Consider then, the path B which is the boundary of R_4 of Example 4. Then Eq. (13) holds since $f(z)$ is single-valued and analytic inside B. Using the directions of integration as indicated in Figure 4. we obtain

$$\int_B f(z)dz = \int_C f(z)dz + \int_\ell f(z)dz + \int_{C_1} f(z)dz + \int_{-\ell} f(z)dz = 0 \quad (14)$$

where B is composed of the separate curves C, ℓ, C_1 and $-\ell$ and the result of Eq. (11) is used.   Hence

$$\int_C f(z)dz \;+\; \int_{C_1} f(z)dz \;=\; 0 \tag{15}$$

since the straight line portions add to zero by Eq. (12). The directions of integration on C and C_1 are indicated in Figure 4. and are very important. Notice

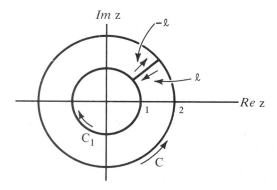

Figure 4.

that the region R_4 is <u>always</u> <u>on</u> <u>the</u> <u>left</u> as you proceed along C or C_1. This is consistent with the counterclockwise direction assumed for closed curves.

Although Eq. (15) was derived assuming an isolated singularity at z = 0, it actually holds when the singularity is at any other point, z = z_0, with C and C_1 being arbitrary smooth curves that enclose z_0. Actually, the result holds even if f(z) is analytic inside both C and C_1, but then both integrals are zero by Eq. (13) and thus Eq. (15) is not necessary. What Eq. (15) says is that the integral around an isolated singular point is the same for all closed paths that surround it (and no other singular points). Rewriting Eq. (15) we obtain

$$\int_C f(z)dz \;=\; -\int_{C_1} f(z)dz$$

$$=\; \int_{-C_1} f(z)dz, \tag{16}$$

where the direction of integration on C_1 has been reversed and indicated by a $-C_1$ (this direction is now the same as on C). Equation (16) is the extension of Eq. (13) when f(z) is single-valued but not analytic at a point interior to the closed curve.

Of course, as it now stands, Eq. (16) must be evaluated as a line integral since it also encloses the singular point of f(z). The usefulness of Eq. (16), at this time, is that a path other than C may be chosen appropriately to simplify the calculation of the given line integral. In the next section we will show how Eq. (16) may be evaluated (since it is not necessarily zero), using Laurent series. By using arguments similar to those above, more than one singular point may be considered, as in Exercise 8.

In the section on differentiation of complex functions it was pointed out that differentiation of analytic functions gave the same results as if z were a real variable. For complex integration the same result holds. That is if F(z) is defined by

$$F(z) \;=\; \int_{z_0}^{z} f(w)dw, \tag{17}$$

where f(w) is analytic and single-valued in a region containing the path joining z_0 and z, then F(z) is an analytic function and

$$F'(z) \;=\; f(z). \tag{18}$$

This is essentially the same as the fundamental theorem of calculus. The extension of Eq. (17) to fixed limits is given by

$$\int_{\alpha}^{\beta} f(z)dz \;=\; \int_{\alpha}^{z_0} f(z)dz \;+\; \int_{z_0}^{\beta} f(z)dz \tag{19}$$

$$=\; \int_{z_0}^{\beta} f(z)dz \;-\; \int_{z_0}^{\alpha} f(z)dz$$

$$=\; F(\beta) \;-\; F(\alpha), \tag{20}$$

assuming f(z) is analytic. This result should be compared to that obtained in Section 5.7 for vector line integrals when the force field was conservative.

Example 5.   Evaluate the integral of Example 1. using Eq. (20).
Since $f(z) \;=\; z^2$ is analytic for all   z   we have

$$\int_{0}^{2+i} z^2 dz \;=\; \frac{z^3}{3}\Bigg|_{0}^{2+i} \;=\; \frac{(2+i)^3}{3} \;=\; 2/3 \;+\; 11i/3$$

because the derivative of $z^3/3$ is z^2 (Eq. (18)).

Exercises

1. Evaluate the integral of Examples 1. and 2. along the straight line segment
    joining the limit points.

2. Evaluate $\displaystyle\int_0^{2+i} (1 + \bar{z}^2)\, dz$ along the paths of Examples 1. and 2.

3. Evaluate $\displaystyle\int_C |z|^2 dz$ and $\displaystyle\int_C z^2 dz$ as line integrals around the closed curve

    which is the square with vertices $(0,0)$, $(1,0)$, $(1,1)$, $(0,1)$. Does Eq. (13)
    hold in both cases?

4. Evaluate $\displaystyle\int_C ze^z dz$ around the ellipse $x^2 + 2y^2 = 1$ using any appropriate

    technique.

5. Using Eq. (20) to evaluate each of the following.

    a) $\displaystyle\int_0^{2+i} (z^2 - 2iz + 1)dz$ on the line joining the points.

    b) $\displaystyle\int_2^{2i} \frac{dz}{z-1}$ on the circular arc $z = 2e^{i\theta}$, $0 \le \theta \le \pi/2$.

    c) $\displaystyle\int_0^{i+1} ze^z dz$ on the line joining the points.

    d) $\displaystyle\int_1^{i} \frac{zdz}{z^2+4}$ on the straight line joining the points.

6. Evaluate $\displaystyle\int_C \frac{4dz}{z+1}$ around the ellipse $2x^2 + y^2 = 4$. (Hint: The circle

    $z + 1 = e^{i\theta}$, $0 \le \theta \le 2\pi$, also encloses the singular point $z = -1$. Use this
    circle and its given parametric representation in Eq. (16) to evaluate the
    given integral.)

7. Let the closed curve C be the square with sides $x = \pm 2$ and $y = \pm 1$.
    Evaluate each of the following.

a) $\int_C \dfrac{dz}{(z-3)^3}$ b) $\int_C \dfrac{dz}{(z-1)}$ c) $\int_C \dfrac{dz}{(z-1)^2}$

     (Hint: Where appropriate use the concepts developed in the hint
            of Exercise 6.)

8. To extend Eq. (15) or Eq. (16) to the case where the closed curve encloses two
    or more isolated singular points consider evaluating

$\int_C \dfrac{dz}{z^2-1}$, where C is the circle $|z| = 3$, by the following steps:

a) Let C_1 and C_2 be given by $|z+1| = 1$ and $|z-1| = 1$ respectively.

b) Join C_1 and C_2 to C by straight lines ℓ_1 and ℓ_2 respectively.

c) Let the closed curve B consist of C, C_1, C_2, ℓ_1, and ℓ_2.

d) Justify that $\int_B \dfrac{dz}{z^2-1} = 0.$

e) Justify that $\int_B \dfrac{dz}{z^2-1} = \int_C \dfrac{dz}{z^2-1} - \int_{C_1} \dfrac{dz}{z^2-1} - \int_{C_2} \dfrac{dz}{z^2-1}$,

     where all integrations are counterclockwise.

f) Evaluate $\int_{C_1} \dfrac{dz}{z^2-1}$ and $\int_{C_2} \dfrac{dz}{z^2-1}$.

     (Hint: $\dfrac{1}{z^2-1} = \dfrac{1}{2}\left(\dfrac{1}{z-1} - \dfrac{1}{z+1}\right)$)

g) What is $\int_C \dfrac{dz}{z^2-1}$ using parts d, e, and f?

9. Evaluate each of the following assuming Leibnitz's rule (Section 3.3) holds.

a) $\dfrac{d}{dw} \displaystyle\int_0^w \dfrac{z\,dz}{z^2+2z+3}$ b) $\dfrac{d}{dw} \displaystyle\int_{-3w}^{w^2} ze^z dz$

c) $\dfrac{d}{dw} \displaystyle\int_1^{2w} \dfrac{dz}{z-w}$

d) $\dfrac{d}{dw} \displaystyle\int_C \dfrac{z\,dz}{z-w}$, C is the circle $|z| = 1$.

10. Use Eqs. (18) and (20) to show that $\displaystyle\int_C \dfrac{dz}{(z-z_0)^n} = 0$ for $n = 2,3,4 \cdots$.

6.6 RESIDUES AND PARTIAL FRACTIONS

In this section we will continue the discussion of complex integration around closed curves. Previously we have seen that if $f(z)$ is analytic and single-valued inside a closed contour C, then

$$\int_C f(z)\,dz = 0. \tag{1}$$

However, in many important cases, $f(z)$ is not analytic at all points inside C. If $f(z)$ has <u>isolated</u> <u>singular</u> <u>points</u> inside C (z_0 is an isolated singular point if $f(z)$ is analytic for $0 < |z-z_0| < \gamma$ but not analytic at z_0), then Eq. (16) and Exercise 8. of Section 6.5 tell us that $\displaystyle\int_C f(z)\,dz$ is the sum of integrals over closed paths going around each of the isolated singularities:

$$\int_C f(z)\,dz = \int_{C_1} f(z)\,dz + \cdots + \int_{C_n} f(z)\,dz, \tag{2}$$

where the closed paths C_k are shown in Figure 1. The direction of integration is indicated and is always counterclockwise.

Equation (2) looks like a lot of work to evaluate. However, using Laurent Series and the fact that $f(z)$ has only isolated singular points, each of the integrals on

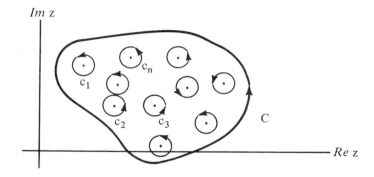

Figure 1.

the right of Eq. (2) is very simple to evaluate. If $f(z)$ has an isolated singularity at z_0, then we saw in Section 6.4 that $f(z)$ will have a Laurent series:

$$f(z) = \sum_{n=0}^{\infty} a_n (z-z_0)^n + \frac{b_1}{(z-z_0)} + \frac{b_2}{(z-z_0)^2} + \cdots \tag{3}$$

for $0 < |z - z_0| < \gamma$. The portion of Eq. (3) with negative powers is called the **principal part** of $f(z)$ and it is this part that will show the character of the singularity. Now if C_1 is a closed curve surrounding z_0 lying inside $0 < |z-z_0| < \gamma$, then

$$\int_{C_1} f(z)dz = \int_{C_1} \sum_{n=0}^{\infty} a_n(z-z_0)^n dz + \int_{C_1} \frac{b_1\,dz}{(z-z_0)} + \int_{C_1} \frac{b_2\,dz}{(z-z_0)^2} + \cdots . \tag{4}$$

The first term on the right of Eq. (4) is zero since Eq. (1) applies in this case. The second term in Eq. (4) yields:

$$\int_{C_1} \frac{b_1\,dz}{(z-z_0)} = b_1 \ln\,(z-z_0) \Bigg|_{C_1} \tag{5}$$

$$= b_1 [\ln|z-z_0| + 2\pi i - \ln|z-z_0| - 0i] \tag{6}$$

$$= 2\pi i b_1 \tag{7}$$

where we have used the fact that if C_1 is a closed contour, the magnitude of z at the two ends of the contour must be the same, while the argument has increased by 2π. In a similar manner (see Exercise 10, Section 6.5).

$$\int_{C_1} \frac{b_n\,dz}{(z-z_0)^n} = 0, \qquad n = 2,3,4, \cdots . \tag{8}$$

Thus, from Eqs. (7) and (8) we get

$$\int_{C_1} f(z)dz = 2\pi i b_1 . \tag{9}$$

The quantity b_1 is called the **residue** of $f(z)$ at the singular point z_0. From Eq. (9) we have

$$b_1 \; = \; \frac{1}{2\pi i} \int_{C_1} f(z)dz, \tag{10}$$

where $f(z)$ has only the isolated singularity at z_0 inside of C_1. The integral around C_1, of course, is done in the counterclockwise direction.

Example 1.   Evaluate $\int_C \frac{z}{z+1} \, dz$ where   C   is the circle $|z+1| = 1$.

By dividing $z + 1$ into z, we obtain

$$\frac{z}{z+1} \; = \; 1 \; - \; \frac{1}{z+1} \; .$$

Thus $b_1 \; = \; -1$ and Eq. (9) yields

$$\int_C \frac{z}{z+1} \, dz \; = \; -2\pi i.$$

Equation (9) covers the situation when $f(z)$ has only one singular point inside C. If $f(z)$ has a finite number of isolated singular points inside C, then Eq. (2) may be used, and each term on the right of Eq. (2) may be evaluated using Eq. (9). Thus

$$\int_C f(z)dz \; = \; 2\pi i \; [k_1 \; + \; k_2 \; + \; \cdots \; + \; k_n], \tag{11}$$

where k_i is the residue at z_i, $i = 1, \cdots$ n.   The z_i, $i = 1, \cdots$ n, are all of the singular points of $f(z)$ inside of C.

Example 2.   Evaluate $\int_C \frac{dz}{z(z+1)}$ when   C   is the circle $|z| = 2$.

The integrand has a singularity at $z = 0$ and $z = -1$. To find the residues at these points we note that

$$\frac{1}{z(z+1)} \; = \; \frac{1}{z} \; - \; \frac{1}{z+1} \; . \tag{12}$$

Thus the residues are 1 and -1 respectively and hence Eq. (11) yields

$$\int_C \frac{dz}{z(z+1)} \; = \; 2\pi i [1 - 1] \; = \; 0.$$

Example 2. was not difficult to evaluate since the residues at 0 and -1 could easily be obtained. However, for more complicated functions the _partial fraction expansion_, Eq. (12) of Example 2, is more difficult to obtain. The partial fraction expansion of a rational function $f(z)$ (the quotient of two polynomials) is the sum of the principal parts of $f(z)$ about each of its isolated singularities. In Eq. (12) for instance the function has singularities at $z = 0$ and $z = -1$. The term $1/z$ is the principal part of $f(z)$ at $z = 0$ ($1/(z+1)$ is analytical at $z = 0$ and thus has a Taylor series expansion, Section 6.4, about $z = 0$) and the term $-1/(z+1)$ is the principal part of $f(z)$ at $z = -1$ ($1/z$ is analytical at $z = -1$ and thus has a Taylor series expansion about $z = -1$). Thus, to find the partial fraction expansion of a function it is necessary to find the principal parts about each of its singularities. To do this $f(z)$ is assumed to have only a finite number of poles of finite order in the complex plane. For a pole or order m at $z = c$ the Laurent series (Section 6.4) for $f(z)$ about $z = c$ is

$$f(z) = \frac{b_m}{(z-c)^m} + \cdots + \frac{b_1}{(z-c)} + \sum_{n=0}^{\infty} a_n (z-c)^n .$$ (13)

Multiplying both sides of Eq. (13) by $(z-c)^m$ yields

$$(z-c)^m f(z) = b_m + b_{m-1} (z-c) + \cdots$$ (14)

and thus

$$b_m = (z-c)^m f(z) \Big|_{z=c}$$ (15)

since all the other terms of Eq. (14) vanish at $z = c$. If Eq. (14) is differentiated once we get

$$\frac{d}{dz} [(z-c)^m f(z)] = b_{m-1} + 2b_{m-2} (z-c) + \cdots$$ (16)

so that

$$b_{m-1} = \frac{d}{dz} [(z-c)^m f(z)] \Big|_{z=c} ,$$ (17)

again since all the other terms of Eq. (16) vanish at $z = c$. Continuing in this fashion we obtain

$$b_{m-n} = \frac{1}{n!} \frac{d^{(n)}}{dz^n} [(z-c)^m f(z)] \Big|_{z=c} .$$ (18)

If m is not too large, Eq. (18) is quite useful for determining the coefficients of the terms with negative powers in Eq. (13). If Eq. (18) is used for each of the singularities of a rational function f(z), then its partial fraction expansion can be obtained.

The following examples will illustrate how to use partial fractions to calculate residues so that integrals around singular points may be evaluated.

Example 3.   Evaluate $\int_C \dfrac{(z-1)dz}{z(z+1)}$ when C is the circle $|z| = 2$.

The partial fraction expansion of the integrand is given by

$$\frac{z-1}{z(z+1)} = \frac{A}{z} + \frac{B}{z+1} \cdot \tag{19}$$

Using Eq. (15) we obtain the values of A and B as

$$A = z \cdot \frac{z-1}{z(z+1)} \Bigg|_{z=0} = -1$$

$$B = (z+1) \cdot \frac{z-1}{z(z+1)} \Bigg|_{z=-1} = 2.$$

Thus the residue at z = 0 is −1 and the residue at z = −1 is 2 so that Eq. (11) yields

$$\int_C \frac{(z-1)dz}{z(z+1)} = 2\pi i[-1 + 2] = 2\pi i.$$

Example 4.   Evaluate $\int_C \dfrac{dz}{z^2(z-1)}$ when C is the circle $|z| = 2$.

In this case the partial fraction expansion is

$$\frac{1}{z^2(z-1)} = \frac{A}{z^2} + \frac{B}{z} + \frac{D}{z-1} , \tag{20}$$

where

$$A = z^2 \cdot \frac{1}{z^2(z-1)} \Bigg|_{z=0} = -1,$$

$$B = \frac{d}{dz}\left(z^2 \cdot \frac{1}{z^2(z-1)}\right)\Bigg|_{z=0} = \frac{-1}{(z-1)^2}\Bigg|_{z=0} = -1$$

and

$$D = (z-1) \cdot \frac{1}{z^2(z-1)} \Bigg|_{z=1} = 1,$$

using Eqs. (15) and (17). Thus

$$\frac{1}{z^2(z-1)} = -\frac{1}{z^2} - \frac{1}{z} + \frac{1}{z-1}$$

and hence the residue at $z = 0$ is -1 while the residue at $z = 1$ is 1. Therefore

$$\int_C \frac{dz}{z^2(z-1)} = 2\pi i \,[-1 + 1] = 0$$

by Eq. (11).

The concept of partial fractions, however, is not applicable when the integrand is not a rational function. In this case, though, the residue may be found by determining the Laurent series using the approach developed in Section 6.4 This is illustrated in the following example.

Example 5.   Evaluate $\displaystyle\int_C \frac{e^z dz}{z^2}$ when   C   is the circle $|z| = 1$.

Using the power series for e^z about $z = 0$ we obtain

$$\frac{e^z}{z^2} = \frac{1 + z + \frac{z^2}{2!} + \cdots + \frac{z^n}{n!} + \cdots}{z^2} .$$

Dividing z^2 into each of the numerator terms thus yields

$$\frac{e^z}{z^2} = \frac{1}{z^2} + \frac{1}{z} + \frac{1}{2!} + \frac{z}{3!} + \cdots$$

as the Laurent series for e^z/z^2. Therefore the residue is   1   and hence

$$\int_C \frac{e^z dz}{z^2} = 2\pi i.$$

The approach for finding the residue of Example 5. could also have been used on the

previous examples. In practice one finds that sometimes partial fractions are used
(if the order of the pole is not too large) and other times the Laurent series are
used. The concept of partial fractions has other applications, for instance, in
the work with Laplace transforms of Chapter 7.

These six sections on complex variables have been a very brief introduction to
the topic. However, through the examples and the exercises, many of the very useful
topics have at least been introduced and illustrated.

Exercises

1. If C is the circle $|z| = 3$ in the counterclockwise sense and if

$$g(z_0) = \int_C \frac{2z^2 + 1}{z - z_0} \, dz,$$

a) find $g(0)$ and $g(1)$.

b) what is $g(z_0)$ if $|z_0| > 3$?

2. Evaluate each of the following integrals.

a) $\displaystyle\int_C \frac{z^2+1}{z(z^2-1)} \, dz$ when C is the circle $|z| = 2$

b) $\displaystyle\int_C \frac{e^{z+1}}{(z+1)^3} \, dz$ when C is the circle $|z-1| = 6$

c) $\displaystyle\int_C \frac{z}{z^2+3z+2} \, dz$ when C is the circle $|z+1| = 1/2$

d) $\displaystyle\int_C \frac{\ln(z+2)}{z^3} \, dz$ when C is the circle $|z| = 1$

e) $\displaystyle\int_C \frac{(z^2-z)}{(z+2)^2(z^2+1)} \, dz$ when C is the circle $|z| = 3$

3. Find the order of the pole and the residue for each singular point for each of the following functions.

a) $\dfrac{1 - e^{-2z}}{z^3}$

b) $\dfrac{1}{z^3 + 3z^2}$

c) $\dfrac{e^z}{(z-1)^2}$

d) $\dfrac{2z + 1}{(z^2-1)^2}$

e) $\dfrac{3-4z}{z^3-3z^2+2z}$

f) $\dfrac{z}{z^4-1}$

4. Find the partial fraction expansion of each of the following functions.

a) $\dfrac{z^2 + 4}{z(z^2+5z+4)}$

b) $\dfrac{1}{z^3(z+1)}$

c) $\dfrac{2z - 1}{z^2(z^2-1)}$

d) $\dfrac{1 - z}{z(z+2)^2}$

5. The partial fraction expansion for $\dfrac{1}{(z+1)(z^2+2z+2)}$

can be obtained as follows:

a) Find A in $\dfrac{1}{(z+1)(z^2+2z+2)} = \dfrac{A}{z+1} + \dfrac{Bz+C}{z^2+2z+2}$ using Eq. (15).

b) Then use $\dfrac{1}{(z+1)(z^2+2z+2)} - \dfrac{A}{z+1} = \dfrac{Bz+C}{z^2+2z+2}$

to find B and C by combining fractions on the left and then equating like coefficients.

c) Check your answers by factoring z^2+2z+2 into complex linear factors and finding the appropriate residue using Eq. (15).

7

LAPLACE TRANSFORMS

7.1 INTRODUCTION TO LAPLACE TRANSFORMS

The Laplace transform provides a very useful mathematical tool for the solution of differential equations and general systems analysis problems that arise in engineering and other applications. It is the intent of this chapter to present the basic fundamentals of the Laplace transform and to illustrate its effectiveness on some problems not previously covered. There are, of course, many other applications in such areas as control systems analysis, linear differential equations with polynomial coefficients, partial differential equations and differential-difference equations, but these will not be covered here.

The Laplace transform of a function $f(t)$ is denoted by $F(s)$ and is defined by the integral:

$$F(s) = \int_0^\infty e^{-st} f(t)dt, \tag{1}$$

provided the integral exists. There are various ways the integral in Eq. (1) may fail to exist. If $f(t)$ is not appropriately behaved near $t = 0$ (for instance $f(t) = 1/t$) or not appropriately behaved for large values of t (for instance $f(t) = e^{t^2}$), then the transform does not exist. In addition, if $f(t)$ has certain types of discontinuities or an infinite number of them for $t > 0$, the integral in Eq. (1) may not exist either. However, as will be shown later, if $f(t)$ has a finite number of "jump" discontinuities, then the integral will exist, and will provide an important extension to material covered in Chapter 1. For most application oriented problems $f(t)$ will be suitably continuous and the integral will exist.

The variable s that appears in Eq. (1) is in general a complex variable:

$$s = \sigma + i\omega . \tag{2}$$

However, if Chapter 6 has not been covered, the reader may still proceed under the assumption that s is real ($\omega = 0$) without any essential loss in generality. In this case the material on the inverse transform appearing in Section 7.5 should be omitted.

An alternate notation used with Laplace transforms is to indicate the Laplace transform of f(t) by L{f(t)} , that is

$$L\{f(t)\} = F(s) = \int_0^\infty e^{-st} f(t) \, dt. \tag{3}$$

The use of the basic definition in Eq. (1) will be illustrated by deriving the Laplace transform of several fundamental functions. The simplest application is to find the transform of the unit constant function:

$$f(t) = 1, \qquad\qquad 0 \leq t . \tag{4}$$

Substituting Eq. (4) into Eq. (1) we obtain:

$$L\{1\} = \int_0^\infty e^{-st} 1 \, dt, \tag{5}$$

$$= \frac{-1 \; e^{-st}}{s} \Bigg|_{t=0}^{t=\infty} \qquad\qquad ' \tag{6}$$

since s is independent of t. The value of the right hand side of Eq. (6) is rigorously found using a limit process which is shown here for illustrative purposes:

$$L\{1\} = \frac{-e^{-st}}{s} \Bigg|_{t=0}^{t=\infty} = \lim_{R \to \infty} \frac{-e^{-st}}{s} \Bigg|_{t=0}^{t=R} \tag{7}$$

$$= \lim_{R \to \infty} \left(\frac{-e^{-sR}}{s} + \frac{1}{s} \right)$$

$$= \frac{1}{s} , \tag{8}$$

provided the real part of s (denoted by Re s) is positive. Equation (8) shows that the Laplace transform of the constant function one exists for all values of s for which Re s > 0.

The Laplace transform of the exponential function is given by

$$L\{e^{at}\} = \frac{1}{s-a} \qquad \text{Re } s > \text{Re } a \quad . \tag{9}$$

This can be derived by the following steps:

$$L\{e^{at}\} = \int_0^\infty e^{-st} e^{at} \, dt \tag{10}$$

$$= \int_0^\infty e^{-(s-a)t} \, dt$$

$$= \lim_{R \to \infty} \left. \frac{-e^{-(s-a)t}}{s-a} \right|_{t=0}^{t=R}$$

$$= \lim_{R \to \infty} \frac{-e^{-(s-a)R}}{s-a} + \frac{1}{s-a} \tag{11}$$

$$= \frac{1}{s-a} \quad , \tag{12}$$

again provided that the Re $(s-a) > 0$. Thus the Laplace transform of the exponential function is given by Eq. (9) and exists for Re $s >$ Re a. These restrictions on Equations (8) and (9), however, don't really hamper the application aspects of their use as in general they can easily be satisfied.

Three other elementary transforms which will be needed in the next section are:

$$L\{t\} = \frac{1}{s^2} \qquad \text{Re } s > 0, \tag{13}$$

$$L\{\cos \omega t\} = \frac{s}{s^2 + \omega^2} \qquad \text{Re } s > 0, \tag{14}$$

and

$$L\{\sin \omega t\} = \frac{\omega}{s^2 + \omega^2} \qquad \text{Re } s > 0, \tag{15}$$

which are left as an exercise to derive. More transforms will be presented in Section 7.3 after Laplace transforms have been applied to the solution of differential equations in Section 7.2

The above equations are not only needed in obtaining the transforms of known

functions but are also needed in obtaining the inverse transform. That is, suppose
the transform of an unknown function is known, then Equations (8), (9) and (13) to
(15) may be used to find the appropriate time domain function. The following example
will illustrate this.

Example 1.   Find $f(t)$ when $F(s) = \dfrac{s}{s^2 + 4}$.

   To find $f(t)$, look at the right hand side of the transforms that have
been presented and note that Eq. (14) is of the same form with $\omega^2 = 4$
(or $\omega = 2$). Thus

$$f(t) = \cos 2t,$$

by referring to the left hand side of Eq. (14).

This procedure for finding the inverse is possible since it can be shown that the
Laplace transform of a given function is unique. The table of transform pairs
presented in Section 7.3 will facilitate the process of finding the inverse transform
when more complicated problems are encountered.

   In most applications of the Laplace transform the linearity property plays a
very important role. It is stated mathematically as:

$$L\{a\ f(t) + b\ g(t)\} \;=\; a\ F(s) + b\ G(s) \tag{16}$$

provided a and b are constants and the transforms of $f(t)$ and $g(t)$ exist. Equation (16) can be derived by simply using the definition as given in Eq. (1) and
Example 2. will illustrate how Eq. (16) is applied.

Example 2.   Find the Laplace transform of $f(t) = 2t - \sin 3t$.
   From Eq. (16) we have

$$L\{f(t)\} \;=\; L\{2t - \sin 3t\}$$

$$= \;L\{2t\} + L\{-\sin 3t\}$$

$$= \;2L\{t\} - L\{\sin 3t\}$$

$$= \;\frac{2}{s^2} - \frac{3}{s^2 + 9}\,,$$

where Eqs. (13) and (15) have been used.

   The final result that is needed for the application of Laplace transforms to
the solution of differential equations is the finding of the transform of the deriva-

tive of a function:

$$L\{\frac{df}{dt}\} \; = \; s \; F(s) \; - \; f(0),$$ (17)

where $F(s)$ is the Laplace transform of $f(t)$ and $f(0)$ is the value of $f(t)$ as t approaches zero. Equation (17) can be derived using the definition of the Laplace transform:

$$L\{\frac{df}{dt}\} \; = \; \int_{0}^{\infty} e^{-st} \frac{df}{dt} \; dt$$

$$= \; e^{-st} \; f(t) \; \Bigg|_{t=0}^{t=\infty} \; + \; s \int_{0}^{\infty} e^{-st} \; f(t) \; dt$$

$$= \; -f(0) \; + \; s \; F(s).$$

The first step is accomplished using integration by parts and the second step is accomplished by recognizing the Laplace transform of $f(t)$ and assuming $e^{-st} f(t)$ approaches zero as t gets large. The Laplace transform of the second derivative of a function can be obtained using Eq. (17) twice:

$$L\{\frac{d^2f}{dt^2}\} \; = \; L\{\frac{d}{dt} \; (\frac{df}{dt})\}$$

$$= \; s \; L\{\frac{df}{dt}\} \; - \; \frac{df}{dt}(0)$$

$$= \; s(s \; F(s) \; - \; f(0)) \; - \; \frac{df}{dt}(0)$$

$$= \; s^2 \; F(s) \; - \; s \; f(0) \; - \; \frac{df}{dt}(0) \; .$$

For higher order derivatives this process can be continued.

Exercises

1. Derive Eq. (13) using integration by parts.

2. Derive Eqs. (14) and (15) using either integration by parts or Eq. (12) with

    $a = i\omega$ and $e^{ix} = \cos x + i \sin x$.

3. Find the Laplace transform of each of the following functions.

a) $6e^{-2t}$ b) $5t + 2 \cos 3t$

c) $2 \dfrac{dy}{dt} + 6y$ d) $3 - 4 \sin \frac{1}{2}t$

e) $\dfrac{d^2 y}{dt^2} + 2 \dfrac{dy}{dt} + y$ f) $e^{-t} + \cos t$

4. Use integration by parts to find the Laplace transform of each of the following functions.

a) t^2 b) te^{at}

5. Find f(t) given the following transforms F(s).

a) $\dfrac{2}{s} - \dfrac{3}{s^2}$ b) $\dfrac{5}{s^2 + 4} - \dfrac{3}{s}$

c) $\dfrac{1}{s(s + 1)}$ (Use partial fractions: $\dfrac{1}{s(s + 1)} = \dfrac{A}{s} + \dfrac{B}{s + 1}$ and solve for A and B).

7.2 LAPLACE TRANSFORM SOLUTIONS OF DIFFERENTIAL EQUATIONS

The most important use of the Laplace transform is in the solution of linear constant coefficient differential equations with initial conditions. In this section several examples will be worked that will illustrate the techniques to be used. Of course the solutions can be checked using methods discussed in Chapter 1.

The advantage of using Laplace transforms to solve the equations is that the differential equations and initial conditions are transformed into an algebraic equation for the transform of the unknown function. Finding the inverse of this transform will then yield the desired solution, usually after using partial fractions to put the expressions in a form to use the transform pairs of the previous section.

Example 1.   Solve

$$\frac{dy}{dt} + 2y = 1 \tag{1}$$

$$y(0) = 2$$

using Laplace transforms.

If the Laplace transform of both sides of the differential equation is taken we get

$$L\{\frac{dy}{dt} + 2y\} = L\{1\} . \tag{2}$$

Using linearity, Eq. (8) and Eq. (17) of the last section we have

$$L\{\frac{dy}{dt}\} + 2 \, L\{y(t)\} = \frac{1}{s}$$

and

$$s \, Y(s) - y(0) + 2 \, Y(s) = \frac{1}{s} , \tag{3}$$

where $Y(s) = L\{y(t)\}$. Equation (3) is an algebraic equation for $Y(s)$, and is solved to give

$$Y(s) = \frac{2s + 1}{s(s + 2)} , \tag{4}$$

since $y(0) = 2$. To find $y(t)$, the right side of Eq. (4) is expanded using partial fractions:

$$\frac{2s + 1}{s(s + 2)} = \frac{A}{s} + \frac{B}{s + 2} \tag{5}$$

$$= \frac{1/2}{s} + \frac{3/2}{s + 2} \tag{6}$$

where A and B have been determined by equating appropriate coefficients in the numerator terms of the left and right sides of Eq. (5) or by using methods of Section 6.6. If Eq. (6) is substituted into Eq. (4) we get

$$Y(s) = \frac{1/2}{s} + \frac{3/2}{s + 2}$$

so that, after referring to the transform pairs of Section 7.1, we find:

$$y(t) = \frac{1}{2} + \frac{3}{2} e^{-2t} .$$

It should be noted that this is exactly the same solution as would have been obtained using methods of Chapter 1.

The next two examples will illustrate the Laplace transform solution of two second order problems. They are chosen in particular to further illustrate the partial fractions that are encountered in this approach.

Example 2.   Solve, using Laplace transforms:

$$\frac{d^2y}{dt^2} + 3\frac{dy}{dt} + 2y = 2e^{-3t}$$

(7)

$$y(0) = 0, \quad \frac{dy}{dt}(0) = 1.$$

Using the same steps as shown in Example 1, we obtain

$$s^2 Y(s) - sy(0) - \frac{dy}{dt}(0) + 3(sY(s) - y(0)) + 2Y(s) = \frac{2}{s+3},$$

which again is an algebraic equation for $Y(s)$. Using the initial conditions and solving for $Y(s)$ yields

$$(s^2 + 3s + 2)Y(s) = 1 + \frac{2}{s+3}$$

(8)

$$= \frac{s+5}{s+3},$$

so that

$$Y(s) = \frac{s+5}{(s+3)(s^2+3s+2)}.$$

(9)

Again, Eq. (9) represents the Laplace transform of the solution and partial fractions are used to find $y(t)$ as follows:

$$\frac{s+5}{(s+3)(s^2+3s+2)} = \frac{A}{s+3} + \frac{B}{s+1} + \frac{C}{s+2}$$

since the denominator has the three linear factors. The coefficients, A, B and C may now be found as before to yield:

$$Y(s) = \frac{1}{s+3} + \frac{2}{s+1} - \frac{3}{s+2}.$$

(10)

From Eq. (10), the solution of Eqs. (7) is found to be

$$y(t) = e^{-3t} + 2e^{-t} - 3e^{-2t}.$$

(11)

Example 3.   Solve, using Laplace transforms

$$\frac{d^2y}{dt^2} + y = \cos 2t$$

(12)

$$y(0) = 0, \quad \frac{dy}{dt}(0) = 1.$$

The solution of Eqs. (12) follows the same pattern as in the previous examples. The Laplace transform yields:

$$s^2 Y(s) - 1 + Y(s) = \frac{s}{s^2 + 4}$$

so that

$$(s^2 + 1) Y(s) = \frac{s^2 + s + 4}{s^2 + 4} \tag{13}$$

and

$$Y(s) = \frac{s^2 + s + 4}{(s^2 + 4)(s^2 + 1)} \ .$$

In this case, the denominator does not have further linear factors with real coefficients. Thus the appropriate partial fraction expansion is

$$\frac{s^2 + s + 4}{(s^2 + 4)(s^2 + 1)} = \frac{As + B}{s^2 + 1} + \frac{Cs + D}{s^2 + 4} \ ,$$

where the coefficients may be found easiest by equating numerator terms.

Thus

$$Y(s) = \frac{s/3 + 1}{s^2 + 1} - \frac{s/3}{s^2 + 4}$$

$$= \frac{s/3}{s^2 + 1} + \frac{1}{s^2 + 1} - \frac{s/3}{s^2 + 4}$$

and hence the solution $y(t)$ is given by

$$y(t) = \frac{1}{3} \cos t + \sin t - \frac{1}{3} \cos 2t \ . \tag{14}$$

The above examples have illustrated the use of Laplace transforms in solving certain differential equations, and, perhaps more importantly, have illustrated the procedure that must be used to find the inverse transform. Before proceeding to a final example, it is worth pointing out some generalizations based on these examples. First, it should be noted that the complete initial value problem is solved as a whole, rather than in the steps of finding the homogeneous and nonhomogeneous solutions and then satisfying the initial condition after the general solution has been found. Of course the transform method does not yield the general solution without further work, if that is what is desired. Secondly, the student should recognize the characteristic equation (as defined in Section 1.2) that appears as the coefficient of Y(s) in the above examples. (See, for instance, Eqs. (8) and (13).)

The usefulness of the Laplace transform technique is perhaps not apparent from the above examples. However, more complicated problems or similar repetitious problems are handled in an easier manner with transforms. For these cases the necessity of partial fractions can be minimized with the use of a more elaborate set of transform pairs which are developed with a given set of problems in mind. In addition the transform technique does provide an advantage for a large variety of problems, as illustrated by Example 4. and the succeeding sections. In these cases the methods of Chapter 1 are either very cumbersome or not appropriate.

The final example to be discussed here is the solution of a system of differential equations using Laplace transforms. Systems of equations can also be solved using vector and matrix methods or simply substitution techniques. However, for system analysis that arise in various control problems in engineering or other applications, the Laplace transform method is particularly appropriate.

Example 4.   Solve

$$\frac{dy}{dt} - y - 2x = 1 \quad , \qquad y(0) = 2$$

$$\frac{dx}{dt} - 2y - x = t \quad , \qquad x(0) = 1$$

(15)

by means of the Laplace transform.

The procedure is essentially the same as that used on the above examples. The Laplace transform is taken of both equations to yield:

$$sY(s) - y(0) - Y(s) - 2X(s) = \frac{1}{s}$$

$$sX(s) - x(0) - 2Y(s) - X(s) = \frac{1}{s^2} \, .$$

(16)

Again the transform has reduced the differential equations to algebraic equations which may be solved for X(s) and Y(s):

$$X(s) \;\; = \;\; \frac{s^3 + 3s^2 + 3s - 1}{s^2(s - 3)(s + 1)}$$

and

$$Y(s) \;\; = \;\; \frac{2s^3 + s^2 - s + 2}{s^2(s - 3)(s + 1)} \, .$$

Each of these may be expanded using partial fractions to yield:

$$X(s) = \frac{1/3}{s^2} - \frac{11/9}{s} + \frac{31/18}{s-3} + \frac{1/2}{s+1}$$

and   (17)

$$Y(s) = \frac{-2/3}{s^2} + \frac{7/9}{s} + \frac{31/18}{s-3} - \frac{1/2}{s+1}$$

so that the solutions of Eqs. (15) are

$$x(t) = \frac{1}{3} t - \frac{11}{9} + \frac{31}{18} e^{3t} + \frac{1}{2} e^{-t}$$

and   (18)

$$y(t) = -\frac{2}{3} t + \frac{7}{9} + \frac{31}{18} e^{3t} - \frac{1}{2} e^{-t}.$$

The format of the solutions, as given in Eqs. (18), is already familiar to the student and the solutions may be verified as being correct by substitution back into Eqs. (15). The partial fraction expansion here illustrates the procedure when a term such as $(s - a)^2$ is encountered in the denominator.

Exercises

1. Use Laplace transforms to solve the following first order differential equations with initial conditions.

a) $\frac{dy}{dt} + 3y = t + 3, \qquad y(0) = 1$ b) $\frac{dy}{dt} + y = \sin 2t, \qquad y(0) = 0$

2. Use Laplace transforms to solve the following second order differential equations with initial conditions.

a) $\frac{d^2y}{dt^2} + 4 \frac{dy}{dt} + 3y = t, \qquad y(0) = 0, \frac{dy}{dt}(0) = 0$

b) $\frac{d^2y}{dt^2} + y = \sin 2t, \qquad y(0) = 1, \frac{dy}{dt}(0) = 0$

c) $\frac{d^2y}{dt^2} + 2 \frac{dy}{dt} + y = 2, \qquad y(0) = 0, \frac{dy}{dt}(0) = 0$

d) $\frac{d^2y}{dt^2} + 5 \frac{dy}{dt} + 6y = 1, \qquad y(0) = 1, \frac{dy}{dt}(0) = 1$

3. Use Laplace transforms to solve the following systems of equations.

a) $\dfrac{dx}{dt} + x + 2y = t$

$\dfrac{dy}{dt} - x + 4y = 1$ $\qquad\qquad$ $x(0) = 1,\ y(0) = 1$

b) $\dfrac{d^2y}{dt^2} - \dfrac{dx}{dt} + 3y = 3t$ $\qquad$ $x(0) = 0,\ \dfrac{dx}{dt}(0) = 0,$

$\dfrac{d^2x}{dt^2} + 2\dfrac{dy}{dt} + 2x = -2$ $\qquad$ $y(0) = 0,\ \dfrac{dy}{dt}(0) = 0$

7.3 FURTHER PROPERTIES OF THE LAPLACE TRANSFORM AND THEIR APPLICATION

In the previous two sections an introduction to Laplace transforms as applied to differential equations has been given. It is the purpose of this and succeeding sections to extend that material to other types of problems than can be handled with the brief material developed so far. Our approach will be to derive some results that allow us to considerably extend the transform pairs we have seen and then apply them to the solution of some problems.

The most important extension for the type of problem considered so far is the Laplace transform of the product of the exponential function and any other function f(t):

$$L\{e^{at} f(t)\} = F(s - a), \tag{1}$$

assuming the transform F(s) of f(t) exists. This result is obtained easily from the definition:

$$L\{e^{at} f(t)\} = \int_0^\infty e^{-st} e^{at} f(t)\,dt$$

$$= \int_0^\infty e^{-(s-a)t} f(t)\,dt$$

$$= F(s - a),$$

since

$$F(s) = \int_0^\infty e^{-st} f(t)\,dt. \tag{2}$$

The result given in Eq. (1) is useful for extending the transform pairs that were obtained in Section 7.1 and also for finding the inverse Laplace transform of certain types of expressions. These are illustrated in the following examples.

Example 1.   Find the Laplace transform of

$$f(t) = e^{-2t} \cos 3t.$$

Since $f(t)$ is the product of an exponential and a cosine function we may use Eq. (1) with $a = -2$ to obtain

$$L\{e^{-2t} \cos 3t\} = \frac{s + 2}{(s+2)^2 + 9} \tag{3}$$

$$= \frac{s + 2}{s^2 + 4s + 13} ,$$

since

$$L\{\cos 3t\} = \frac{s}{s^2 + 9} \tag{4}$$

from Eq. (14) of Section 7.1.

Example 2.   Find the inverse Laplace transform of

$$F(s) = \frac{4}{s^2 + 2s + 2} . \tag{5}$$

Since the denominator cannot be factored into real linear factors, the partial fraction method of the previous sections will not work. Thus, we complete the square in the denominator to obtain:

$$F(s) = \frac{4}{(s+1)^2 + 1} . \tag{6}$$

This is of the form of Eq. (15), Section 7.1, with s replaced by $s + 1$ and $\omega = 1$. Thus Eq. (6) implies that

$$f(t) = 4e^{-t} \sin t . \tag{7}$$

Further examples of the use of Eq. (1) will be shown when additional differential equations are solved at the end of this section.

Some other results that are needed for a variety of applications will now be derived. The first one is obtained by taking the derivative of the integral appearing in Eq. (2), the definition of the Laplace transform. Thus we have

$$\frac{dF}{ds} = \int_0^\infty \frac{\partial e^{-st}}{\partial s} f(t)dt, \tag{8}$$

$$= \int_0^\infty e^{-st}(-t)f(t)dt. \tag{9}$$

The differentiation in Eq. (8) can be justified whenever the transform in Eq. (2) exists. Equation (9) can be rewritten as

$$L\{-tf(t)\} = \frac{dF}{ds} \ . \tag{10}$$

Further differentiations of Eq. (9) would then yield the general result:

$$L\{(-t)^n f(t)\} = \frac{d^n F}{ds^n} \ . \tag{11}$$

Equations (10) and (11) are needed not only for finding the Laplace transform of such functions as t sin 2t, but also in solving differential equations in which polynomial coefficients appear.

A second result that is useful in various applications, particularly where a change of variables is encountered, is given by

$$L\{f(ct)\} = \frac{1}{c} F(\frac{s}{c}), \qquad c > 0. \tag{12}$$

This result can also be obtained from the definition of the transform with a change in variable of integration:

$$L\{f(ct)\} = \int_0^\infty e^{-st} f(ct)dt$$

$$= \int_0^\infty e^{-s\tau/c} f(\tau) \frac{1}{c} d\tau,$$

where $\tau = ct$. Thus

$$L\{f(ct)\} = \frac{1}{c} \int_0^\infty e^{-(s/c)\tau} f(\tau)d\tau \tag{13}$$

since c is a constant. Equation (13) yields Eq. (12) as long as c is positive (otherwise the existence of the integral as τ approaches infinity is affected).

Example 3.   Given that

$$L\{f(t)\} = \frac{s}{s^2 + 4}$$

find $L\{f(3t)\}$.

Applying Eq. (12) we have

$$L\{f(3t)\} = \frac{1}{3} \frac{s/3}{(s/3)^2 + 4}$$

$$= \frac{s}{s^2 + 36} \ .$$

In some applications, control problems and systems analysis problems in particular, it is desirable to determine the initial value and final value for a function f(t) directly from its transform F(s) without first finding the inverse transform. This is possible through the following two results:

$$\lim_{t \to 0} f(t) = \lim_{s \to \infty} sF(s), \tag{14}$$

which yields the initial value of f(t), and

$$\lim_{t \to \infty} f(t) = \lim_{s \to 0} sF(s), \tag{15}$$

which yields the final value of f(t). Neither of these will be derived here, but it should be pointed out that Eq. (15) is not valid when F(s) has poles (Section 6.4) for Re s > 0 or s pure imaginary. In these cases it is not difficult to show that the final value does not exist.

Example 4.   Find the initial and final value of f(t) when

$$F(s) = \frac{6}{s(s^2 + s + 2)} \ .$$

Using Eq. (14) we have

$$\lim_{t \to 0} f(t) = \lim_{s \to \infty} s \frac{6}{s(s^2 + s + 2)}$$

$$= \lim_{s \to \infty} \frac{6}{s^2 + s + 2}$$

$$= 0 \quad,$$

as the initial value. Likewise Eq. (15) gives

$$\lim_{t \to \infty} f(t) = \lim_{s \to 0} s \frac{6}{s(s^2 + s + 2)}$$

$$= \lim_{s \to 0} \frac{6}{s^2 + s + 2}$$

$$= 3 \quad,$$

as the final value.

In both cases of the example, the desired value could also be obtained by finding the inverse transform.

Table I is presented for easy reference in the use of Laplace transforms. The table presents the transform pairs that have been developed in Section 7.1 as well as others that can be obtained easily with the results of this section and that are common in many applications. The use of the table will enable the reader to quickly locate either the transform of a time function f(t) or the inverse transform of F(s). The following two examples illustrate the use of Table I in solving differential equations. Neither of these equations can be handled by the techniques in Section 7.1.

Example 5.   Using Laplace transforms, solve the initial
                        value problem

$$\frac{d^2 y}{dt^2} + 2 \frac{dy}{dt} + 2y = 4$$

$$y(0) = 0, \quad \frac{dy}{dt}(0) = 0 \quad.$$

Proceeding by the same steps as used in Section 7.2 we find that the transform of the solution is given by

$$Y(s) = \frac{4}{s(s^2 + 2s + 2)} \quad.$$

Table I

Laplace Transform Pairs

$f(t)$ for $0 \leq t$		$F(s) = \int_0^\infty e^{-st} f(t)dt$	
1.	1	$\dfrac{1}{s}$	$s > 0$
2.	t	$\dfrac{1}{s^2}$	$s > 0$
3.	t^n for $n > -1$	$\dfrac{n!}{s^{n+1}}$	$s > 0$
4.	e^{at}	$\dfrac{1}{s - a}$	$s > a$
5.	$t\,e^{at}$	$\dfrac{1}{(s-a)^2}$	$s > a$
6.	$\cos \omega t$	$\dfrac{s}{s^2 + \omega^2}$	$s > 0$
7.	$\sin \omega t$	$\dfrac{\omega}{s^2 + \omega^2}$	$s > 0$
8.	$e^{at} \sin \omega t$	$\dfrac{\omega}{(s-a)^2 + \omega^2}$	$s > a$
9.	$e^{at} \cos \omega t$	$\dfrac{s-a}{(s-a)^2 + \omega^2}$	$s > a$
10.	$a\,f(t) + b\,g(t)$	$a\,F(s) + b\,G(s)$	
11.	$\dfrac{df}{dt}$	$s\,F(s) - f(0)$	
12.	$\dfrac{d^2f}{dt^2}$	$s^2\,F(s) - s\,f(0) - \dfrac{df}{dt}(0)$	
13.	$(-t)^n\,f(t)$	$\dfrac{d^n F}{ds^n}$	
14.	$f(ct)$	$\dfrac{1}{c}F(\dfrac{s}{c})$	$c > 0$

Since the denominator cannot be factored into real linear factors, the
appropriate partial fraction expansion is

$$\frac{4}{s(s^2 + 2s + 2)} = \frac{A}{s} + \frac{Bs + C}{s^2 + 2s + 2} \ ,$$

where it is found that A = 2, B = -2 and C = -4.   Thus

$$Y(s) = \frac{2}{s} - \frac{2s + 4}{s^2 + 2s + 2}$$

$$= \frac{2}{s} - \frac{2s + 4}{(s+1)^2 + 1}$$

$$= \frac{2}{s} - \frac{2(s + 1)}{(s+1)^2 + 1} - \frac{2}{(s+1)^2 + 1} \ ,$$

where the denominator has been rewritten by completing the square and
the constant in the numerator has been broken up to form the appropriate
s + 1 factor in the numerator.   Referring to pairs 1, 8, and 9 of Table I
we obtain

$$y(t) = 2 - 2e^{-t} \cos t - 2e^{-t} \sin t$$

as the desired solution.

Example 6.   Using Laplace transforms, solve the initial
                          value problem

$$\frac{d^2 y}{dt^2} + 4y = \cos 2t$$

$$y(0) = 0, \ \frac{dy}{dt}(0) = 0 \ .$$

As before, we find that

$$Y(s) = \frac{s}{(s^2 + 4)^2}$$

is the Laplace transform of the solution.   Y(s) cannot be further expanded
using partial fractions and a look at Table I does not immediately yield
y(t).   However, it is noted that Y(s) is of the form of the derivative

of the transform of sin 2t, and hence if we use the result 13, with n = 1, along with the transform pair 7 of Table I we obtain

$$y(t) = \frac{1}{4} t \sin 2t$$

as the desired solution.

Exercises

1. Use Eq. (1) to find the Laplace transform of each of the following functions.

    a) $3te^{-2t}$

    b) $e^{at} \sin \omega t$

    c) $t^4 e^{-3t}$

    d) $e^{at} \cos \omega t$

2. Use Eq. (11) to find the Laplace transform of each of the following functions.

    a) $2te^{-t}$

    b) $t \cos \omega t$

    c) $t \sin \omega t$

    d) $t^3 e^{-2t}$

3. Use Eq. (12) to show that the inverse transform of $F(cs + d)$ is $\frac{1}{c} e^{-dt/c} f(\frac{t}{c})$ when c > 0.

4. Use the results of Exercise 3. to find the inverse Laplace transform of each of the following functions.

    a) $\dfrac{3s + 1}{9s^2 + 6s + 5}$

    b) $\dfrac{3}{(5s + 2)^3}$

5. Find the initial and final values of the time functions having the following Laplace transforms.

    a) $\dfrac{3}{s(2s + 3)}$

    b) $\dfrac{1}{s(s^2 + 3s + 6)}$

    c) $\dfrac{2s^2 + 1}{s(s^2 + s + 1)}$

    d) $\dfrac{s + 3}{s(s^2 + 4)}$

6. Solve the following initial value problems using Laplace transforms.

a) $\dfrac{d^2 y}{dt^2} + 4 \dfrac{dy}{dt} + 8y = 2,$ $y(0) = 0,$ $\dfrac{dy}{dt}(0) = 0$

b) $\dfrac{d^2 y}{dt^2} + 2 \dfrac{dy}{dt} + 2y = 5 \cos t,$ $y(0) = 1,$ $\dfrac{dy}{dt}(0) = 0$

c) $\dfrac{d^2 y}{dt^2} + 9y = \sin \omega t,$ $y(0) = 0,$ $\dfrac{dy}{dt}(0) = 0$

    (Solve for $\omega^2 \neq 9$ and then for $\omega^2 = 9$)

d) $\dfrac{d^2 y}{dt^2} + 2 \dfrac{dy}{dt} + y = 4,$ $y(0) = 2,$ $\dfrac{dy}{dt}(0) = 0$

7.4   STEP FUNCTIONS AND IMPULSE FUNCTIONS

In the solution of constant coefficient differential equations in the last two sections, the use of Laplace transforms offers no particular advantage over the methods of Chapter 1. However, if the forcing function is not continuous, then the methods of Chapter 1 are either very cumbersome or impossible and the use of Laplace transforms becomes imperative. In this section we will develop the necessary trans- form material and then apply it to solve some appropriate problems.

The discussion of Section 7.1 pointed out that the Laplace transform does exist for functions with certain types of discontinuities. The simplest of these are known as <u>jump</u> <u>discontinuities</u>. A function f(t) is said to have a jump discontinuity at a point c if the left and right limits of f(t) at   c   are finite but are not equal. The unit step function

$$u(t,c) \; = \; \begin{cases} 0 & t < c \\ 1 & t \geq c \end{cases} \qquad c \geq 0 \qquad\qquad (1)$$

is an example of a function with a jump discontinuity at t = c. Notice that the right limit of u(t,c) at t = c is 1, while the left limit of u(t,c) at t = c is 0. The graph of u(t,c) is shown in Figure 1, where we see that the graph of u(t,c) is a smooth curve except for the finite break occuring at the point   c.   The constant forcing function that we have encountered in our previous work is obtained from u(t,c) by setting c = 0 in Eq. (1) and multiplying by the appropriate constant. This relationship occurs as long as the initial time is t = 0.

The unit step function as given in Eq. (1) can be used in various ways to create

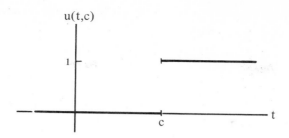

Figure 1.

other functions that have jump discontinuities. For instance,

$$g(t) \ = \ u(t,1) \ - \ 2u(t,2) \tag{2}$$

has jump discontinuities at t = 1 and t = 2 as shown in Figure 2. The unit step

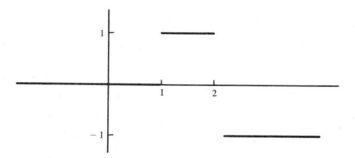

Figure 2.

function can also be used to translate a function to the right. Suppose f(t) is
defined for t ≥ 0, then the function

$$g(t) \ = \ u(t,c) \ f(t-c) \tag{3}$$

represents f(t) shifted to the right (since c ≥ 0) c units. This is shown in Figure 3.

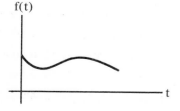

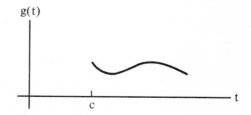

Figure 3.

where it can be seen that g(t) must be zero for t < c, and the functional values of
g(t) for t > 0 are obtained from f(t) by a shift to the right of c units. The
translated function, as shown in Eq. (3), is particularly important in applications
involving delays or lags. In these cases various effects on the physical systems
do not start initially and must be accounted for through functions of the type shown
in Eq. (3).

The Laplace transform of the unit step function and the translated function
can be obtained readily from the definition given in Section 7.1. For the translated function we have

$$L\{u(t,c)f(t-c)\} \; = \; \int_{0}^{\infty} e^{-st} u(t,c)f(t-c)dt$$

$$= \; \int_{c}^{\infty} e^{-st} f(t-c)dt$$

since u(t,c) is zero for $0 \leq t < c$. By changing the variable of integration to
$\tau = t - c$, we obtain

$$L\{u(t,c)f(t-c)\} \; = \; \int_{0}^{\infty} e^{-s(\tau+c)} f(\tau)d\tau$$

$$= \; e^{-cs} \int_{0}^{\infty} e^{-s\tau} f(\tau)d\tau$$

since e^{-cs} is independent of the integration variable τ. Thus

$$L\{u(t,c)f(t-c)\} \; = \; e^{-cs} F(s), \tag{5}$$

assuming the transform F(s) exists. Equation (5) may now be used to find the transform of the unit step function, since we may set f(t-c) = 1 in Eq. (5) to obtain

$$L\{u(t,c)\} \; = \; e^{-cs} L\{1\}$$

$$= \; \frac{e^{-cs}}{s} \; , \; s > 0. \tag{6}$$

The transform pairs given in Eqs. (5) and (6) can be added to those already appearing
in Table I of Section 7.3. Example 1. and 2. show their use in determining transforms
and inverse transforms.

Example 1.   Find the Laplace transform of

$$h(t) = \begin{cases} 2 & 0 \le t < 1 \\ 2 + t & 1 \le t \end{cases} . \tag{7}$$

In order to use the above material, we note first that

$$h(t) = 2 + u(t,1)t. \tag{8}$$

However, the second term on the right of Eq. (8) is not yet of the correct form to use Eq. (5). To obtain the proper form we must add and subtract the correct terms as shown:

$$h(t) = 2 + u(t,1)\{(t - 1) + 1\}$$

$$= 2 + u(t,1)(t - 1) + u(t,1). \tag{9}$$

The Laplace transform of h(t) may now be found by using Eqs. (5) and (6) to yield:

$$L\{h(t)\} = \frac{2}{s} + \frac{e^{-s}}{s^2} + \frac{e^{-s}}{s} \tag{10}$$

since the f(t) of Eq. (5) is given by f(t) = t for this example.

Example 2.   Find the inverse Laplace transform of

$$H(s) = \frac{3 - 2e^{-3s}}{s^2} . \tag{11}$$

Rewriting H(s) as follows

$$H(s) = \frac{3}{s^2} - \frac{2e^{-3s}}{s^2} ,$$

we find that

$$h(t) = 3t - 2u(t,3)(t - 3) \tag{12}$$

by reference to Eq. (5). Note that f(t) = t so that f(t - 3) = t - 3 in the use of Eq. (5). The h(t) as given in Eq. (12) may be written as

$$h(t) = \begin{cases} 3t & 0 \leq t < 3 \\ t + 6 & 3 \leq t \end{cases} \qquad . \tag{13}$$

The function $h(t)$ of Example 1. and the function $h'(t)$ found from Eq. (13) both exhibit a jump discontinuity and would be typical of functions appearing in applications where discontinuities or delays in actions might occur.

A second type of discontinuity that can be handled with Laplace transforms is that exhibited by the impulse function. Technically the impulse function is not a function in the mathematical sense and thus it is difficult to write it mathematically. For our purposes we will simply say that the impulse function is an idealization of a pulse function that has a very large amplitude acting over a very short time duration. Hence, we will derive its Laplace transform using the previously developed unit step function and then take the limiting case as the duration of the pulse becomes zero. To this end then, we define

$$g(t) = \begin{cases} 0 & t < 0 \\ \dfrac{1}{A} & 0 \leq t < A \\ 0 & A \leq t \end{cases} \qquad , \tag{14}$$

which is known as a pulse of height $\dfrac{1}{A}$ and duration of A units. Notice that the area under the pulse is exactly one unit, and hence it is a unit pulse. Rewriting $g(t)$ in terms of the unit step function we have

$$g(t) = \frac{1}{A} - \frac{1}{A} u(t,A) , \qquad t \geq 0 \tag{15}$$

and hence its transform is given by

$$G(s) = \frac{1}{As} - \frac{e^{-As}}{As}$$

$$= \frac{1}{s} \frac{1 - e^{-As}}{A} \qquad .$$

Now the unit impulse function is obtained from $g(t)$ by letting A approach zero. Assuming the limiting process can also be done on the transform, we then obtain the transform of the impulse function from the following:

$$\lim_{A \to 0} G(s) = \lim_{A \to 0} \frac{1}{s} \frac{1 - e^{-As}}{A}$$

(16)

$$= \lim_{A \to 0} \frac{1}{s} \frac{se^{-As}}{1} = 1,$$

by using L'Hospital's rule.

The unit impulse function is usually denoted as $\delta(t)$ and is known as the Dirac delta function. Thus Eq. (16) can be written as

$$L\{\delta(t)\} = 1.$$

(17)

From Eq. (14) and the limiting process used to obtain Eq. (16) we see that $\delta(t)$ is an impulse occuring at time $t = 0$. The impulse function for a positive value of t is obtained by translating $\delta(t)$:

$$h(t) = u(t,c) \, \delta(t - c).$$

(18)

However, since $\delta(t)$ is zero except for $t = 0$, the step function in Eq. (18) is redundant. Thus $\delta(t - c)$ is used to denote the unit impulse at $t = c$ and its transform is given by

$$L\{\delta(t - c)\} = e^{-cs}.$$

(19)

The following example will illustrate the use of the above material in the solution of a differential equation involving an impulse function.

Example 3.   Solve the initial value problem

$$\frac{d^2y}{dt^2} + y = 2\delta(t - \pi)$$

(20)

$$y(0) = 1, \quad \frac{dy}{dt}(0) = 0.$$

In this case we must use Laplace transforms to obtain

$$(s^2 + 1)Y(s) - s = 2e^{-\pi s}$$

or

$$Y(s) \; = \; \frac{s}{s^2 + 1} \; + \; \frac{2e^{-\pi s}}{s^2 + 1} \; .$$

Thus

$$y(t) \; = \; \cos t \, + \, 2u(t,\pi)\sin(t - \pi), \qquad\qquad (21)$$

by referring to Table I of Section 7.3 and using Eq. (5).

It is illustrative to take a closer look at the solution as shown in Eq. (21). It can be rewritten in the form

$$y(t) \; = \; \begin{cases} \cos t & 0 \le t < \pi \\ \cos t - 2 \sin t & \pi \le t, \end{cases} \qquad\qquad (22)$$

from which it can be seen that for the interval $0 \le t < \pi$ the solution is as expected from solving the initial value problem of Eq. (21) with no forcing function. Then at time $t = \pi$, when the impulse is imparted on the system, the solution changes to account for the impulse. Since $\sin \pi$ is zero, the solution as shown in Eq. (21) or Eq. (22) is continuous at $t = \pi$. However, its derivative will not be continuous at $t = \pi$, which is to be expected, since the impulse function is not continuous at that point and the discontinuities of the two sides of the equations are then balanced.

Exercises

1. Write each of the following functions in terms of the unit step function and then find its Laplace transform.

a) $\; f(t) \; = \; \begin{cases} 0 & 0 \le t < 1 \\ 2 & 1 \le t < 3 \\ -1 & 3 \le t \end{cases}$ b) $\; f(t) \; = \; \begin{cases} 0 & 0 \le t < 1 \\ (t - 4)^2 & 4 \le t \end{cases}$

c) $\; f(t) \; = \; \begin{cases} 1 & 0 \le t < 1 \\ t^2 & 1 \le t \end{cases}$ d) $\; f(t) \; = \; \begin{cases} 0 & 0 \le t < \pi/2 \\ \sin t & \pi/2 \le t \end{cases}$

e) $\; f(t) \; = \; \begin{cases} 0 & 0 \le t < 1 \\ e^{-t/3} & 1 \le t \end{cases}$ f) $\; f(t) \; = \; \begin{cases} t & 0 \le t < 1 \\ 1 & 1 \le t \end{cases}$

2. Find the inverse Laplace transform of each of the following functions.

a) $\dfrac{1 - e^{-2s}}{s}$

b) $\dfrac{e^{-s}}{s + 3}$

c) $\dfrac{e^{-\pi s}}{s^2 + 4}$

d) $\dfrac{e^{-s}}{s(s + 2)}$

3. Find the solution of each of the following initial value problems.

a) $\dfrac{d^2 y}{dt^2} + 2 \dfrac{dy}{dt} + 2 y = 1 - u(t,1)$, $\qquad y(0) = 0$, $\dfrac{dy}{dt}(0) = 0$

b) $\dfrac{d^2 y}{dt^2} + y = h(t)$, $\qquad y(0) = 1$, $\dfrac{dy}{dt}(0) = 0$,

$$\text{where } h(t) = \begin{cases} 0 & 0 \le t < \pi/2 \\ 1 & \pi/2 \le t < \pi \\ 0 & \pi \le t \end{cases}$$

c) $\dfrac{d^2 y}{dt^2} + 3 \dfrac{dy}{dt} + 2 y = u(t,2)$, $\qquad y(0) = 1$, $\dfrac{dy}{dt}(0) = 0$

d) $\dfrac{d^2 y}{dt^2} + y = f(t)$, $\qquad y(0) = 0$, $\dfrac{dy}{dt}(0) = 1$,

where $f(t)$ is given in Exercise 1f).

4. Solve each of the following initial value problems.

a) $\dfrac{d^2 y}{dt^2} + y = 1 + \delta(t - \pi/2)$, $\qquad y(0) = 0$, $\dfrac{dy}{dt}(0) = 0$

b) $\dfrac{d^2 y}{dt^2} + 2 \dfrac{dy}{dt} + 2y = \delta(t - \pi)$, $\qquad y(0) = 0$, $\dfrac{dy}{dt}(0) = 0$

c) $\dfrac{d^2 y}{dt^2} + 3 \dfrac{dy}{dt} + 2y = \delta(t - 2)$, $\qquad y(0) = 1$, $\dfrac{dy}{dt}(0) = 0$

5. Compare the Laplace transform of the solution to the following two problems.

$$\frac{d^2y}{dt^2} + a\frac{dy}{dt} + by = \delta(t), \qquad y(0) = 0, \quad \frac{dy}{dt}(0) = 0$$

and

$$\frac{d^2y}{dt^2} + a\frac{dy}{dt} + by = 0, \qquad y(0) = 0, \quad \frac{dy}{dt}(0) = 1$$

7.5 THE CONVOLUTION INTEGRAL AND THE INVERSE LAPLACE TRANSFORM

The preceding sections have developed the background that is needed for the use of Laplace transforms in working with problems that are encountered in many application areas. This final section on Laplace transforms will present a brief introduction of two concepts, the convolution integral and the inverse transform, that are needed to extend the previous material when more complicated problems arise.

The convolution integral enables one to find the inverse Laplace transform of functions that are written as products without resorting to the use of partial fractions. If H(s) is the Laplace transform of h(t) and G(s) is the Laplace transform of g(t), then the inverse Laplace transform of the product F(s) = H(s) G(s) can be written as

$$f(t) = \int_0^t h(t - \tau) \, g(\tau)d\tau \; . \tag{1}$$

The integral appearing in Eq. (1) is known as the convolution of h(t) and g(t). Before Eq. (1) is derived, several examples will be presented which will illustrate its use and some properties.

Example 1.   Find the inverse Laplace transform of

$$F(s) = \frac{2}{s^2(s^2 + 9)} \; . \tag{2}$$

Instead of employing partial fractions, as was done previously, we note that we can set $H(s) = \dfrac{2}{s^2}$ and $G(s) = \dfrac{1}{s^2 + 9}$, and thus Eq. (1) yields

$$f(t) = \int_0^t 2(t - \tau) \frac{1}{3} \sin(3\tau)d\tau, \tag{3}$$

since $h(t) = 2t$ and $g(t) = \frac{1}{3} \sin 3t$. The function f(t), as shown in

Eq. (3), is the desired function represented as an integral. In this case we can integrate by parts to obtain

$$f(t) = 2(t - \tau)(-\frac{1}{9} \cos 3\tau)\Big|_{\tau=0}^{\tau=t} - \frac{2}{9} \int_0^t \cos(3\tau)\,d\tau$$

$$= \frac{2}{9} t - \frac{2}{27} \sin 3\tau \Big|_{\tau=0}^{\tau=t}$$

$$= \frac{2}{9} t - \frac{2}{27} \sin 3t \ . \tag{4}$$

The integration by parts employed in finding f(t) in Example 1. is generally required when the convolution integral is used to find the inverse transform.

The astute reader might ask at this point whether it makes a difference if the roles of H(s) and G(s) were reversed. That is, in Example 1, would f(t) still be given by Eq. (4) if $H(s) = \frac{2}{s}$ and $G(s) = \frac{1}{s^2 + 9}$. The answer is in the affirmative and can be shown by letting w = t - τ in Eq. (1) to obtain:

$$f(t) = \int_t^0 h(w)\ g(t - w)(-dw)$$

$$= \int_0^t g(t - w)\ h(w)\,dw \ . \tag{5}$$

Comparing Eq. (5) and Eq. (1) it is seen that it does not make any difference which term of the product involved in F(s) is called H(s) and which term is called G(s).

Example 2.   Find the inverse Laplace transform of

$$F(s) = \frac{2}{s^2(s^2 + 9)}$$

by choosing $H(s) = \frac{1}{s^2 + 9}$ and $G(s) = \frac{2}{s^2}$.

In this case we use Eq. (1) with g(t) = 2t and $h(t) = \frac{1}{3} \sin 3t$ to obtain

$$f(t) = \int_0^t (2\tau) \frac{1}{3} \sin 3(t - \tau)\,d\tau.$$

Integration by parts then yields

$$f(t) = 2\tau \frac{1}{9} \cos 3(t - \tau) \Big|_{\tau=0}^{\tau=t} - \frac{2}{9} \int_0^t \cos 3(t - \tau)d\tau$$

$$= \frac{2}{9} t + \frac{2}{27} \sin 3(t - \tau) \Big|_{\tau=0}^{\tau=t}$$

$$= \frac{2}{9} t - \frac{2}{27} \sin 3t, \tag{6}$$

which is the same solution as obtained in Example 1.

In many cases the appropriate choice for H(s) and G(s) will save some computation in the evaluation of the integral.

Turning now to the derivation of Eq. (1), we note that the definition of the Laplace transform says that

$$H(s) = \int_0^\infty e^{-s\eta}h(\eta)d\eta$$

and

$$G(s) = \int_0^\infty e^{-s\tau}g(\tau)d\tau.$$

Thus

$$H(s) \ G(s) = \int_0^\infty e^{-s\eta}h(\eta)d\eta \int_0^\infty e^{-s\tau}g(\tau)d\tau$$

$$= \int_0^\infty \int_0^\infty e^{-s(\eta+\tau)}h(\eta)g(\tau) \ d\eta d\tau, \tag{7}$$

by the definition of the iterated integral, as given in Section 4.1. If we define a new variable of integration $t = \eta + \tau$, for fixed τ, Eq. (7) then yields

$$H(s)\ G(s)\ =\ \int_0^\infty \int_\tau^\infty e^{-st}h(t-\tau)g(\tau)\ dt d\tau$$

$$=\ \int_0^\infty \int_0^t e^{-st}h(t-\tau)g(\tau)\ d\tau dt\ , \tag{8}$$

where the order of integration has been reversed over the "triangular" region shown in Figure 1. Since the e^{-st} term is independent of τ, it can be taken outside the

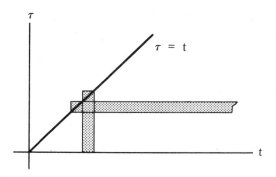

Figure 1.

integral involving the τ integration. Thus

$$H(s)\ G(s)\ =\ \int_0^\infty e^{-st} \int_0^t h(t-\tau)g(\tau)\ d\tau dt$$

$$=\ \int_0^\infty e^{-st}f(t)\ dt, \tag{9}$$

where

$$f(t)\ =\ \int_0^t h(t-\tau)g(\tau)\ d\tau, \tag{10}$$

which is the same as Eq. (1). What Eq. (9) says is that the Laplace transform of $f(t)$ is the product $H(s)G(s)$, and hence $f(t)$ is the inverse Laplace transform of $H(s)G(s)$. This completes the derivation of the result stated in Eq. (1).

The convolution integral is basically a result for finding the inverse Laplace transform when the appropriate products are recognized. The converse is also applicable when functions are defined in the form of Eq. (10), as shown in Example 3.

Example 3.   Find the Laplace transform of the function

$$f(t) = 2 \int_0^t (t - \tau) \sin 3\tau \, d\tau. \tag{11}$$

For the given function we recognize that $h(t) = 2t$ and $g(t) = \sin 3t$, and thus

$$L\{f(t)\} = H(s) \, G(s)$$

by referring to Eqs. (9) and (10). Supplying the correct terms for $H(s)$ and $G(s)$ we then obtain

$$L\{f(t)\} = \frac{2}{s^2} \cdot \frac{3}{s^2 + 9}$$

$$= \frac{6}{s^2(s^2 + 9)}$$

as the Laplace transform of the function defined by Eq. (11).

It should be noted that this type of application is quite specialized in that the integral defining $f(t)$ has to be of the appropriate form. The example, however, does point out again how the transform pair may be used both for finding the transform as well as the inverse transform.

Example 4. illustrates the use of the convolution integral to obtain a result that has importance in a variety of applications.

Example 4.   Find the solution of the initial value problem

$$\frac{d^2 y}{dt^2} + 4y = f(t) \tag{12}$$

$$y(0) = 0, \qquad \frac{dy}{dt}(0) = 0.$$

for functions $f(t)$ whose Laplace transform exists.

Taking the Laplace transform of both sides of Eq. (12) and solving for $Y(s)$ we obtain

$$Y(s) = \frac{F(s)}{s^2 + 4} .$$

Using the convolution integral, we then obtain

$$y(t) = \frac{1}{2} \int_0^t \sin 2(t - \tau) \, f(\tau) d\tau \tag{13}$$

as the solution of the original initial value problem.

Here again we find an integral representation of the solution, but in this case no further information can be obtained until $f(t)$ is known. In this sense Eq. (13) represents the solution to a whole class of problems, one for each $f(t)$. In fact, Eq. (13) represents the solution to Eqs. (12) even in cases where $L\{f(t)\}$ might not exist but the integral in Eq. (13) does exist. For the above example, the function $\sin 2(t - \tau)$ is known as the one-sided <u>Green's function</u>, which depends on the left side of the differential equation and the initial conditions, but not on $f(t)$.

Throughout this chapter on Laplace transforms we have used the concept of the inverse Laplace transform without rigorously defining it. What we have done is to exploit the uniqueness of the transform, and thus if we recognize the function $F(s)$ as the transform of the time function $f(t)$, then $f(t)$ must be the inverse Laplace transform of $F(s)$. In cases where the transformed function is not recognizable, then the mathematical definition is needed in order to proceed with the determination of the time function from the transformed function.

The definition of the Laplace transform has been stated previously as

$$F(s) = \int_0^\infty e^{-st} f(t) dt. \tag{14}$$

When $F(s)$ is obtained from $f(t)$ by the use of Eq. (14), then it can be shown that $f(t)$ can be obtained from $F(s)$ by:

$$f(t) = \frac{1}{2\pi i} \int_{c-i\infty}^{c+i\infty} e^{st} F(s) ds, \tag{15}$$

where c is a real number such that the vertical path of integration lies to the right of all poles of $F(s)$. Equation (15) is the inverse Laplace transform and the two equations together form a transform pair.

In order to evaluate the complex integral appearing in Eq. (15) a finite contour integral is formed, which is related to the given integral, and an appropriate limiting process is used. For positive values of t, $(t > 0)$, as long as $F(s)$ is single-valued, the contour shown in Figure 2. is appropriate. The semicircular contour C_1 (C_1 is given by $|z - c| = R$ for $\pi/2 < \arg(z - c) < 3\pi/2$) is large enough

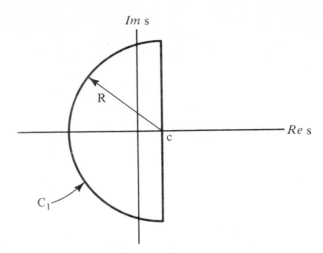

Figure 2.

so that all poles of $F(s)$ are inside of the closed contour. Thus the evaluation of Eq. (15) is given by:

$$f(t) = \lim_{R \to \infty} \frac{1}{2\pi i} \int_{c-iR}^{c+iR} e^{st} F(s) ds. \tag{16}$$

The finite line integral may be obtained from the equation

$$\int_{c-iR}^{c+iR} e^{st} F(s) ds + \int_{C_1} e^{st} F(s) ds = 2\pi i \sum_{j=1}^{n} k_j(t), \tag{17}$$

where, as in Section 6.6, the $k_j(t)$ are the residues of $e^{st} F(s)$ at the poles of $F(s)$. For many functions appearing in applications it is true that

$$\lim_{s \to \infty} |s|^{1+\delta} |F(s)| = 0 \quad \text{for } \delta > 0 \text{ and } \mathrm{Re}(s) \leq c, \tag{18}$$

in which case it is possible to show that the integral along C_1 approaches zero as $R \to \infty$ (for $t > 0$). Thus Eqs. (16) and (17) yield

$$f(t) = \sum_{j=1}^{n} k_j(t), \quad t > 0 \tag{19}$$

as the inverse Laplace transform of $F(s)$.

Example 5.   Find the inverse Laplace transform of

$$F(s) = \frac{1}{s^2(s + 2)}$$

using the above approach.

Since

$$\lim_{s\to\infty} \left| \frac{s^{1+\delta}}{s^2(s + 2)} \right| = 0 , \qquad \delta > 0$$

and since $F(s)$ is single-valued we may use Eq. (19) to give

$$f(t) = \sum \text{Residues of } \frac{e^{st}}{s^2(s + 2)}$$

$$= \frac{e^{st}}{s^2} \bigg|_{s=-2} + \left(\frac{te^{st}}{s + 2} - \frac{e^{st}}{(s + 2)^2} \right) \bigg|_{s=0}$$

where the first term is the residue at $s = -2$ and the second term is the residue at the double pole $s = 0$. Hence the inverse Laplace transform of $F(s)$ is

$$f(t) = \frac{e^{-2t}}{4} + \frac{t}{2} - \frac{1}{4}, \qquad t > 0.$$

It should be pointed out that the vertical path of integration appearing in Eq. (15) must lie to the right of the imaginary axis for Example 5. This was not needed explicitly for Example 5. since we were able to verify that Eq. (19) held, and hence we could use that directly.

The above general results and example hold for $t > 0$. If $t < 0$, then the contour of Figure 2. is no longer appropriate. In this case, to close the contour, a semicircular arc is chosen in the right half plane and it is then necessary to show that the integral along it vanishes as $R\to\infty$. If this happens, then Eq. (19) becomes $f(t) = 0$ for $t < 0$, since $F(s)$ has no poles to the right of the vertical line $s = c$.

Even though Eq. (19) holds for many functions $F(s)$ appearing in applications, there are important exceptions. For instance if $F(s) = 1$, then Eq. (18) no longer holds and the result given in Eq. (19) is therefore not valid. In this case a rigorous analysis of Eq. (16) is needed to yield the inverse transform of $F(s) = 1$, which of course is $f(t) = \delta(t)$. The condition of Eq. (18) is satisfied by $F(s) = s^{-3/2}$, but in this case the contour shown in Figure 2. is not appropriate

since F(s) is not single-valued. In this case a contour such as that shown in
Figure 3. must be chosen. Although the analysis of both of these exceptions is
beyond the scope of this text, they do serve to illustrate the exceptions to the
use of Eq. (19) and what is involved.

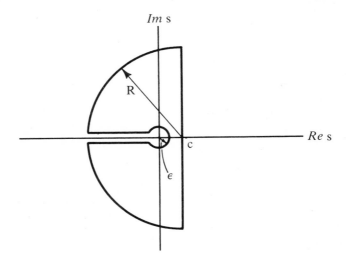

Figure 3.

Exercises

1. Find the inverse Laplace transform of each of the following functions by using
the convolution theorem.

a) $F(s) = \dfrac{1}{s(s + 3)}$

b) $F(s) = \dfrac{1}{s^3(s^2 + 4)}$

c) $F(s) = \dfrac{s}{(s + 2)(s^2 + 1)}$

d) $F(s) = \dfrac{s + 2}{s(s^2 + 2s + 2)}$

2. Find the Laplace transform of each of the following functions.

a) $f(t) = \displaystyle\int_0^t e^{-2(t-u)} \cos(3u)\,du$

b) $f(t) = \displaystyle\int_0^t (t-u)^2 u \, \sin(u)\,du$

3. Find the solution to the initial value problem

$$\frac{d^2 y}{dt^2} + 9y = t, \qquad y(0) = 0, \qquad \frac{dy}{dt}(0) = 0,$$

using the convolution integral.

4. Find the solution of the initial value problem

$$\frac{d^2y}{dt^2} + 2\frac{dy}{dt} + 2y = f(t), \qquad y(0) = 0, \qquad \frac{dy}{dt}(0) = 0,$$

in terms of a convolution integral involving $f(t)$.

5. The equation

$$y(t) + \int_0^t k(t - \tau)y(\tau)d\tau = f(t)$$

in which $k(t)$ and $f(t)$ are known functions is called an _integral_ _equation_ since the unknown function $y(t)$ appears under the integral. Use the convolution integral concept to find $Y(s)$ in terms of $F(s)$ and $K(s)$.

6. Use the result of Exercise 5. to solve each of the following integral equations.

a) $y(t) + \int_0^t (t - \tau) y(\tau)d\tau = t$

b) $y(t) + 2 \int_0^t (t - \tau) y(\tau)d\tau = \cos 3t$

c) $y(t) + \int_0^t \sin 2(t - \tau) y(\tau)d\tau = 1$

7. Find the inverse Laplace transform of each of the following functions using the approach presented in this section. Consider for each case where the vertical path of integration must be chosen.

a) $F(s) = \dfrac{1}{s(s + 1)}$

b) $F(s) = \dfrac{10}{(s + 3)^3}$

c) $F(s) = \dfrac{1}{(s + 1)(s^2 + 4)}$

d) $F(s) = \dfrac{s + 3}{s(s + 2)(s - 1)}$

8

FOURIER SERIES
AND INTEGRALS

8.1 INTRODUCTION TO FOURIER SERIES AND PERIODICITY

The necessity of representing a function $f(x)$ as an infinite sum of sine and cosine terms occurs frequently in many application problems of engineering, physics, economics and other areas. In this case we would have

$$f(x) = \frac{a_0}{2} + \sum_{n=1}^{\infty} a_n \cos nx + b_n \sin nx ,\qquad (1)$$

where the a_n and b_n are constant coefficients and the constant term $a_0/2$ represents $\cos nx$ for $n = 0$. The constant term is written as $a_0/2$ for later convenience when the appropriate formulas are derived for the coefficients. Under certain general conditions a series of the form of Eq. (1), which converges to $f(x)$, is said to be the Fourier series for $f(x)$. The concept of convergence of the series shown in Eq. (1) is no different than that encountered in power series of elementary calculus or in Section 6.4 with complex series since, if a fixed value of x is used, the right hand side becomes a series of constants and thus the previous concepts are appropriate here also. At this time, though, it is not clear that such an expression as given in Eq. (1) is useful or even possible. It is the purpose of this and the following sections to give a brief discussion of these and other questions concerning Fourier series and integrals.

Before it is possible to discuss how to calculate the coefficients a_n and b_n and therefore consider convergence of the series in Eq. (1) it is necessary to consider certain properties of the sine and cosine functions. The first is the concept of periodicity. An arbitrary function $f(x)$ is said to be periodic with period P if

$$f(x + P) = f(x) \qquad (2)$$

for all values of x.

Example 1.   The constant function, $f(x) = A$, is periodic with
                      arbitrary period since

$$f(x + P) \ = \ A \ = \ f(x)$$

    for all values of x.

The functions sin nx and cos nx are periodic, with period 2π.

Example 2.   Show that sin nx is periodic with period 2π for each
                      $n = 1,2,3, \cdots$.

    To do this we must show Eq. (2) is satisfied for $f(x) = \sin nx$ and $P = 2\pi$.
    Thus

$$
\begin{aligned}
f(x + 2\pi) \ &= \ \sin n(x + 2\pi) \\
&= \ \sin(nx + 2n\pi) \\
&= \ \sin nx \cos 2n\pi + \cos nx \sin 2n\pi,
\end{aligned}
\tag{3}
$$

    where the formula for the sine of the sum of two angles have been used.
    Since $\cos 2n\pi = 1$ and $\sin 2n\pi = 0$ for $n = 1,2,3, \cdots$, Eq. (3) reduces to

$$f(x + 2\pi) \ = \ \sin nx \ = \ f(x), \tag{4}$$

    for all x. Hence sin nx is periodic with period 2π.

A similar argument could be given for cos nx and is left as an exercise.
    Actually, each of the functions sin nx for $n = 1,2,3, \cdots$ has a different
fundamental period. The fundamental period is the smallest value of P for which
Eq. (2) holds. For instance sin 3x has fundamental period $2\pi/3$. This can be understood from the fact that if $f(x)$ is periodic with period P, then it is also periodic
with period nP, $n = 1,2,3, \cdots$ (negative periods are not usually encountered in
applications, although they do exist mathematically). This can be seen from the
following:

$$f(x + nP) \ = \ f(x + (n - 1)P + P) \tag{5}$$

$$= \ f(x + (n - 1)P) \tag{6}$$

where Eq. (6) is obtained from Eq. (5) using Eq. (2) with x replaced by x + (n-1)P.
This process can be continued n times so that finally

$$f(x + nP) \;=\; f(x)$$

and thus if f(x) has period P then it also has period nP. It can be seen, then,
that sin nx and cos nx have fundamental periods $2\pi/n$ for each n.

   Returning now to Eq. (1), we see that all terms on the right are periodic with
period 2π. Since the sum of functions with the same period P is also periodic
with period P, we must conclude that if the series on the right of Eq. (1) converges
to f(x), then f(x) must also be periodic with period 2π. One question that this
raises is what other periodic functions are there other than the functions already
considered? To help answer this question we may consider the graph of sin x, shown
in Figure 1. The concept of periodicity tells us that the graph of sin x (or any

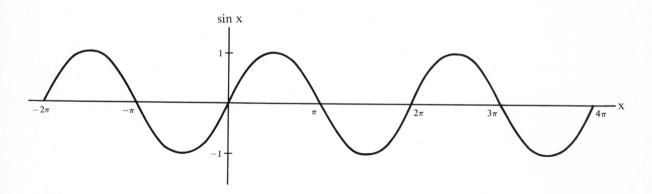

Figure 1.

periodic function f(x)) "repeats" itself every 2π interval (every P interval for
f(x)). Thus, in order to find other periodic functions all we have to do is define
a function for an interval equal to the desired period and then "extend" this function so that we satisfy the periodicity definition of Eq. (2). The following
examples will illustrate this procedure.

Example 3.   To obtain the <u>square</u> <u>wave</u> of period 2π shown in Figure 2.
    consider the function

$$g(x) \;=\; \begin{cases} -1 & -\pi < x \le 0 \\ 1 & 0 < x < \pi. \end{cases} \qquad (7)$$

This is defined for an interval of length 2π and may be extended to other

x values by

$$f(x) \; = \; g(x) \qquad\qquad -\pi < x < \pi \qquad\qquad (8)$$

and

$$f(x + 2\pi) \; = \; f(x), \qquad \text{for all other } x. \qquad\qquad (9)$$

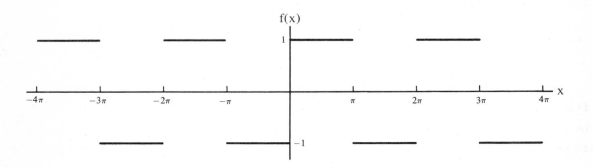

Figure 2.

Equation (9) is the extension of the function of Eq. (8) defined on the interval
$-\pi < x < \pi$. Note that for $f(x)$ in Eq. (8) to be defined x must be in the interval
$(-\pi,\pi)$ and thus $x + 2\pi$ lies in the interval $(\pi, \, 3\pi)$ and hence Eq. (9) has extended
$f(x)$ from $(-\pi,\pi)$ to $(\pi, \, 3\pi)$. The process may now be repeated for x in the interval
$(\pi, \, 3\pi)$ so that $x + 2\pi$ lies in the interval $(3\pi, \, 5\pi)$. Equation (9) also extends
Eq. (8) in the negative direction by considering x in the interval $(-3\pi, \, -\pi)$ and
thus $x + 2\pi$ is in the interval $(-\pi,\pi)$ and thus Eq. (8) holds.

The procedure outlined above and in Example 3. can be carried out for functions
defined over intervals of length other than 2π as shown in the next example.

Example 4.   To obtain the <u>saw tooth</u> wave of period 2 shown in Figure 3.
consider the function

$$g(x) \; = \; x \qquad\qquad -1 < x < 1. \qquad\qquad (10)$$

This is defined for an interval of length 2 and hence may be extended
to other values by

$$f(x) \; = \; x \qquad\qquad -2 < x < 2 \qquad\qquad (11)$$

and

$$f(x + 2) \; = \; f(x) \qquad \text{for all other } x. \qquad\qquad (12)$$

As these two examples have shown, then, we may obtain any number of periodic functions
by taking a function defined over a finite interval (the length of which is the
period) and extending it appropriately.

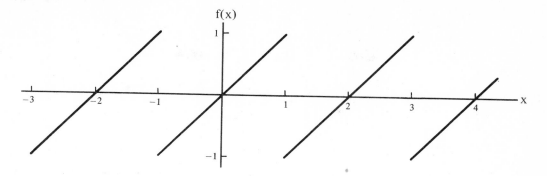

Figure 3.

We now turn to the problem of obtaining the formulas for calculating the coefficients in the series of Eq. (1). There are a variety of ways of doing this, but
the most straightforward is to multiply both sides of Eq. (1) by either cos mx or
sin mx and integrate each term. (The termwise integration of an infinite series is
not always possible. At this point we are assuming it can be done at least for some
functions f(x) so that we can derive the formulas for the coefficients. Later we
will give conditions on f(x) to insure that the termwise integration is permissible.)
The termwise integrations involved can be easily accomplished with the aid of the
following relations, called <u>orthogonality</u> relations:

$$\int_{-\pi}^{\pi} \sin(nx)dx = 0 \tag{13}$$

$$\int_{-\pi}^{\pi} \cos(nx)dx = \begin{cases} 0 & n \neq 0 \\ 2\pi & n = 0 \end{cases} \tag{14}$$

$$\int_{-\pi}^{\pi} \sin(nx)\cos(mx)dx = 0 \qquad \text{all } n, m \tag{15}$$

$$\int_{-\pi}^{\pi} \sin(nx)\sin(mx)dx = \begin{cases} 0 & n \neq m \\ \pi & n = m \end{cases} \tag{16}$$

$$\int_{-\pi}^{\pi} \cos(nx)\cos(mx)dx = \begin{cases} 0 & n \neq m \\ \pi & n = m, \end{cases} \tag{17}$$

where n and m run over zero and all positive integers. Equations (13) to (17) can all be obtained by integrating the given functions, although certain indentities can be used (as shown in Exercise 5.) to simplify the calculations. Orthogonality here does not mean perpendicular in the geometric sense.

Now we are ready to find the coefficients in Eq. (1). First multiply both sides by 1 (cos(0x) = 1) and integrate from $-\pi$ to π (remember that if Eq. (1) is to converge to f(x), f(x) must be given for an interval of length 2π and be the periodic extension for other x) to obtain

$$\int_{-\pi}^{\pi} f(x)dx = \frac{a_0}{2} \int_{-\pi}^{\pi} 1 \, dx + \sum \int_{-\pi}^{\pi} (a_n \cos nx + b_n \sin nx)dx. \qquad (18)$$

Equations (13) and (14) tell us that each term of the infinite sum (denoted simply as $\sum$) is zero and thus the right side of Eq. (18) reduces to the first term only:

$$\int_{-\pi}^{\pi} f(x)dx = \frac{a_0}{2} (2\pi).$$

Solving for a_0 we obtain

$$a_0 = \frac{1}{\pi} \int_{-\pi}^{\pi} f(x)dx, \qquad (19)$$

as the formula for a_0. Equation (19) says that $a_0/2$ represents the average value of f(x) over one period (2π).

To find the a_n, n = 1,2,3 $\cdots$, multiply each side of Eq. (1) by cos mx (m = 1,2,3 $\cdots$) and integrate from $-\pi$ to π to obtain

$$\int_{-\pi}^{\pi} f(x)\cos(mx)dx = \frac{a_0}{2} \int_{-\pi}^{\pi} \cos(mx)dx + \sum \int_{-\pi}^{\pi} a_n \cos(nx)\cos(mx)dx +$$

$$+ \sum \int_{-\pi}^{\pi} b_n \sin(nx)\cos(mx)dx \ . \qquad (20)$$

The sums in Eq. (20) are over n (n = 1,2,3 $\cdots$) while m is held fixed. The first term on the right of Eq. (20) vanishes by Eq. (14) and the last sum vanishes since all terms are zero by Eq. (15). Finally, by Eq. (17), the middle sum has all zero terms except for the one term where n = m so that

$$\int_{-\pi}^{\pi} f(x)\cos(mx)dx \;=\; a_m \int_{-\pi}^{\pi} \cos^2(mx)dx$$

$$=\; \pi\, a_m. \tag{21}$$

Hence

$$a_m \;=\; \frac{1}{\pi} \int_{-\pi}^{\pi} f(x)\cos(mx)dx, \qquad m = 1,2,3 \cdots \tag{22}$$

is the formula for the coefficients a_m for each m. Note, however, that Eq. (22) is the same as Eq. (19) when $m = 0$. It is for this reason that the constant term of Eq. (1) is written as $a_0/2$. In a similar fashion it is found that

$$b_m \;=\; \frac{1}{\pi} \int_{-\pi}^{\pi} f(x)\sin(mx)dx, \qquad m = 1,2,3 \cdots \; . \tag{23}$$

Example 5.   Find the coefficients a_n and b_n for the saw tooth wave of period 2π.

From Example 4. we see that the saw tooth wave of period 2π is given by

$$f(x) \;=\; x \qquad\qquad -\pi < x < \pi$$

and

$$f(x + 2\pi) = f(x) \qquad\qquad \text{all other } x. \tag{24}$$

Thus the a_n are given by substituting in Eq. (22) to yield (with m replaced by n)

$$a_n \;=\; \frac{1}{\pi} \int_{-\pi}^{\pi} x\,\cos(nx)dx$$

$$=\; \frac{1}{\pi} \left(\frac{\cos nx}{n^2} + \frac{x \sin nx}{n} \right) \Bigg|_{-\pi}^{\pi} \;=\; 0 \qquad (n \neq 0),$$

and

$$a_0 \;=\; \frac{1}{\pi} \int_{-\pi}^{\pi} x\,dx \;=\; \frac{1}{\pi}\,\frac{x^2}{2}\, \Bigg|_{-\pi}^{\pi} \;=\; 0.$$

Likewise the b_n are given by Eq. (23) which yields:

$$b_n = \frac{1}{\pi} \int_{-\pi}^{\pi} x \sin(nx)dx$$

$$= \frac{1}{\pi} \left(\frac{\sin nx}{n^2} - \frac{x \cos nx}{n} \right) \Bigg|_{-\pi}^{\pi}$$

$$= \frac{-2\pi \cos n\pi}{n\pi}$$

$$= \frac{2}{n} (-1)^{n+1} .$$

The integrals involved in calculating both a_n and b_n may be obtained by integration by parts or referral to a table of integrals. In evaluating the integrated functions at the end points it is necessary to recall that $\sin n\pi$ is zero for all integers n and that $\cos n\pi$ is 1 for even integers and -1 for odd integers. Finally, if it is assumed that Eq. (1) converges for the saw tooth wave, we then obtain the following representation for $f(x)$ as given in Eqs. (24)

$$f(x) = \sum_{n=1}^{\infty} \frac{2(-1)^{n+1}}{n} \sin nx$$

$$= 2 \left(\sin x - \frac{1}{2} \sin 2x + \frac{1}{3} \sin 3x - \cdots \right).$$

At this point we have shown that if the series of Eq. (1) is to represent a function $f(x)$ we must consider only periodic functions of period 2π and the coefficients a_n and b_n must be given by the formulas

$$a_n = \frac{1}{\pi} \int_{-\pi}^{\pi} f(x)\cos(nx)dx \qquad\qquad (25)$$

and

$$b_n = \frac{1}{\pi} \int_{-\pi}^{\pi} f(x)\sin(nx)dx \ . \tag{26}$$

When the a_n and b_n are given by Eqs. (25) and (26) then Eq. (1) is called the Fourier series of $f(x)$ and the coefficients are called the Fourier coefficients. Since we have not yet considered convergence (and not all series converge even if a_n and b_n are given by Eqs. (25) and (26)) Eq. (1) should be rewritten as

$$f(x) \sim \frac{a_0}{2} + \sum_{n=1}^{\infty} a_n \cos nx + b_n \sin nx \tag{27}$$

to indicate the correspondence of $f(x)$ and its Fourier series.

In the next section we will consider Fourier series for functions with arbitrary periods and for functions not defined over a whole period, then in Section 8.3 we will study convergence of Fourier series.

Exercises

1. Show that $\cos nx$ is periodic with period 2π for each integer n.

2. Show that $\cos nx$ and $\sin nx$ each have fundamental periods $2\pi/n$. Does $f(x) = 1$ have a fundamental period?

3. Determine whether the following functions are periodic and if so, determine the fundamental period.

    a) $\sin 3\pi x$

    b) e^x

    c) $\cos(2\pi x/3)$

    d) $\tan 3x$

    e) $3x^2 + \pi$

    f) $\sin 2x + \cos 3x$

4. Sketch, for at least three periods, the graphs of each of the following functions.

    a) $f(x) = \begin{cases} -1/2 & -\pi < x < 0 \\ 3/2 & 0 < x < \pi \end{cases}$, $f(x + 2\pi) = f(x)$

    b) $f(x) = \begin{cases} -2 & -2 < x < -1 \\ 1 & -1 < x < 1 \\ 1/2 & 1 < x < 2 \end{cases}$, $f(x + 4) = f(x)$

c) $f(x) = |x|$ $\qquad$ $-\pi < x < \pi$, $f(x + 2\pi) = f(x)$

d) $f(x) = \sin x$ $\qquad$ $0 < x < \pi$, $f(x + \pi) = f(x)$

5. Verify the orthogonality relations. For Eqs. (15) to (17) recall that the product of two trigonometric functions can be written as a sum of two other trigonometric functions. The appropriate identity for Eq. (16), for instance is $\sin(nx)\sin(mx) = \frac{1}{2}[\cos(m - n)x - \cos(m + n)x]$.

6. Derive the formulas for b_m as given in Eq. (23).

7. If $f(x)$ is periodic with period P, show that

$$\int_{-P/2}^{P/2} f(x)dx = \int_{a}^{a + P} f(x)dx$$

for any a. Thus the integral of a periodic function over a complete period is the same, no matter what interval is chosen.

8. If $f(x)$ and $g(x)$ are periodic with period P show that

a) $af(x) + bg(x)$ is periodic with period P for any constants a and b.

b) $f(x) g(x)$ is periodic with period P.

c) $\dfrac{df}{dx}$ is periodic with period P.

d) $\displaystyle\int_{0}^{x} f(t)dt$ $\qquad$ may not be periodic.

9. Find the Fourier series representation for each of the following functions.

a) $f(x)$ of Example 3.
b) $f(x)$ of Exercise 4c. This is known as the triangular wave.

c) $f(x) = \begin{cases} 0 & -\pi < x < 0 \\ 1 & 0 < x < \pi, \end{cases}$ $\qquad$ $f(x + 2\pi) = f(x)$

d) $f(x) = \begin{cases} 0 & -\pi < x < 0 \\ x^2 & 0 < x < \pi, \end{cases}$ $\qquad f(x + 2\pi) = f(x)$

e) $f(x) = \sin 3x \qquad -\pi < x < \pi, \qquad f(x + 2\pi) = f(x)$

10. Use the identities $\cos \theta = (e^{i\theta} + e^{-i\theta})/2$ and $\sin \theta = (e^{i\theta} - e^{-i\theta})/2i$ (which are obtainable from Euler's formula of Section 1.2) to show that the series

$$f(x) = a_0 + \sum_{n=1}^{\infty} a_n \cos nx + b_n \sin nx$$

can be written as $f(x) = \sum_{n=-\infty}^{\infty} c_n e^{inx}$, where $c_0 = a_0$, $c_n = \frac{1}{2}(a_n - i b_n)$

and $c_{-n} = \frac{1}{2}(a_n + i b_n)$ for $n = 1,2,3 \cdots$. This is the complex form for a Fourier series and is particularly useful in physics and electrical engineering. Show also that

$$c_n = \frac{1}{2\pi} \int_{-\pi}^{\pi} f(x) e^{-inx} dx.$$

8.2 ARBITRARY PERIODS AND EXTENSIONS

In the last section we introduced the concepts of Fourier series through periodic functions of period 2π. The functions $f(x)$ were basically defined on the interval $(-\pi, \pi)$ and then extended outside of that region by the relation $f(x + 2\pi) = f(x)$. However, the definition of periodicity allows for functions of period other than 2π and hence the problem of representing functions of period $2a$ by Fourier series arises in a natural way.

To find the Fourier series for a periodic function of period $2a$, we begin with the Fourier series for a function of period 2π as found in the last section:

$$g(y) = \frac{a_0}{2} + \sum_{n=1}^{\infty} a_n \cos ny + b_n \sin ny, \qquad (1)$$

where

$$a_n = \frac{1}{\pi} \int_{-\pi}^{\pi} g(y)\cos(ny)dy \tag{2}$$

and

$$b_n = \frac{1}{\pi} \int_{-\pi}^{\pi} g(y)\sin(ny)dy. \tag{3}$$

Now, if we make the change of variable $x = ay/\pi$, which maps the interval $-\pi < y < \pi$ onto the interval $-a < x < a$, then $y = \pi x/a$ and Eqs. (1), (2) and (3) become

$$g(\frac{\pi x}{a}) = f(x) = \frac{a_0}{2} + \sum_{n=1}^{\infty} a_n \cos \frac{n\pi x}{a} + b_n \sin \frac{n\pi x}{a} \tag{4}$$

where

$$a_n = \frac{1}{a} \int_{-a}^{a} f(x)\cos \frac{n\pi x}{a} dx \tag{5}$$

and

$$b_n = \frac{1}{a} \int_{-a}^{a} f(x)\sin \frac{n\pi x}{a} dx. \tag{6}$$

Note that we have renamed the function $g(\frac{\pi x}{a})$ as simply $f(x)$. What these last equations tell us is that if $f(x)$ is defined on $-a < x < a$, then its Fourier series is given by Eq. (4) when the a_n and b_n are found by Eqs. (5) and (6). Since $\cos \frac{n\pi x}{a}$ and $\sin \frac{n\pi x}{a}$ are periodic with period $2a$ for each n, then the $f(x)$ defined on $-a < x < a$ must be extended by $f(x + 2a) = f(x)$ for other x if Eq. (4) is to represent $f(x)$ for all x. If $a = \pi$, then Eqs. (4), (5) and (6) reduce to the first three equations.

Example 1.   Find the Fourier series representation for the function
          $f(x) = |x|$, $-2 < x < 2$ and $f(x + 4) = f(x)$ for all
          other x.

     Notice that $f(x)$ is periodic of period 4 by definition and thus $a = 2$.
     Hence the coefficients are given by

$$a_0 = \frac{1}{2} \int_{-2}^{2} |x| \, dx$$

$$= \frac{1}{2} \int_{-2}^{0} (-x)\,dx + \frac{1}{2} \int_{0}^{2} x\,dx = 2,$$

so that $a_0/2$ is 1, which is the average value of the given function for $-2 < x < 2$. Likewise

$$a_n = \frac{1}{2} \int_{-2}^{2} |x| \cos \frac{n\pi x}{2}\,dx$$

$$= \frac{1}{2} \int_{-2}^{0} -x \cos \frac{n\pi x}{2}\,dx + \frac{1}{2} \int_{0}^{2} x \cos \frac{n\pi x}{2}\,dx$$

$$= -\frac{1}{2} \left[\frac{4 \cos \frac{n\pi x}{2}}{n^2 \pi^2} + \frac{2x \sin \frac{n\pi x}{2}}{n\pi} \right]_{-2}^{0} +$$

$$+ \frac{1}{2} \left[\frac{4 \cos \frac{n\pi x}{2}}{n^2 \pi^2} + \frac{2x \sin \frac{n\pi x}{2}}{n\pi} \right]_{0}^{2}$$

$$= \frac{4 \cos n\pi}{n^2 \pi^2} - \frac{4}{n^2 \pi^2}.$$

Thus

$$a_n = \begin{cases} 0 & \text{if } n \text{ is even} \\[2mm] \dfrac{-8}{n^2 \pi^2} & \text{if } n \text{ is odd.} \end{cases}$$

A similar calculation will show that $b_n = 0$ for all n and hence the Fourier representation for $f(x)$ is

$$f(x) = 1 - \frac{8}{\pi^2} \sum_{m=1}^{\infty} \frac{\cos \frac{(2m-1)\pi x}{2}}{(2m-1)^2}, \tag{7}$$

where the odd integer n has been denoted as $2m - 1$, which is odd for each integer m.

As a check on the calculations for the function of Example 1. we can set x = 1 on the right hand side of Eq. (7). Since $\cos(2m-1)\frac{\pi}{2} = 0$ for all m, we see that the series adds up to zero and hence f(1) = 1, which agrees with the original definition of f(x). Also, since it can be shown that

$$\sum_{m=1}^{\infty} \frac{1}{(2m-1)^2} = \frac{\pi^2}{8} , \tag{8}$$

we have f(2) = 2 and f(0) = 0 from Eq. (7) when x = 2 and 0 respectively. Actually the sum given in Eq. (8) can be found using Eq. (7), once it has been proved that the series in Eq. (7) does indeed converge to f(x).

Up to this point the Fourier series representation for periodic functions have always included both the sine and cosine terms in the general series. However, as Example 5. of Section 8.1 and Example 1. of this section show, certain functions, at least, have Fourier series representations that involve only sine or only cosine terms. In addition, many times in applications it is necessary to represent a function in terms of sines or cosines only. This type representation is possible if one considers f(x) to be defined for only a portion of the basic interval of length 2a, as will be seen in the following.

In order to find series representations involving only one of the trigonometric functions it is necessary to introduce the concepts of <u>even</u> and <u>odd</u> <u>functions</u>. A function f(x) is said to be even if

$$f(x) = f(-x) \tag{9}$$

for all values of x. On the other hand f(x) is said to be odd if

$$f(-x) = -f(x) \tag{10}$$

for all values of x. An even function is symmetrical about the y axis while an odd function is antisymmetrical about the y axis. A typical graph for each is shown in Figure 1. Some even and odd functions are shown in the following example. The student should verify that Eq. (9) or Eq. (10) is appropriately satisfied.

Example 2.   The functions

$$\cos 2x, \ |x|, \ |\sin x|, \ 1, \ x^2$$

are all even functions while the functions

$$\sin 2x, \ \tan x, \ x, \ x^3, \ x^5 - x^3 + x$$

are all odd functions. It is not hard to show that any even power of x will be even and any odd power of x will be odd.

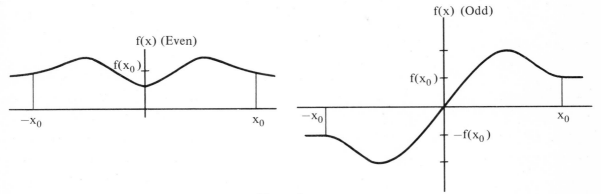

Figure 1.

Not all functions, however, fit into the even or odd category. For instance the commonally encountered functions e^x and $x^2 + x + 1$ are neither even nor odd.

The important properties of even and odd functions that we will need are symbollically written here for reference.

$$
\begin{array}{rcl}
\text{even} + \text{even} &=& \text{even} \\
\text{odd} + \text{odd} &=& \text{odd} \\
\text{even} + \text{odd} &=& \text{neither} \\
\text{even} \times \text{odd} &=& \text{odd} \\
\text{even} \times \text{even} &=& \text{even} \\
\text{odd} \times \text{odd} &=& \text{even}
\end{array} \tag{11}
$$

Most important are the integral relations, which say that the integral of an odd function over a symmetric interval is zero:

$$\int_{-a}^{a} \text{odd} \ dx \ = \ 0 \tag{12}$$

while the integral of an even function over a symmetric interval is twice the integral over the right half of the interval:

$$\int_{-a}^{a} \text{even} \ dx \ = \ 2 \int_{0}^{a} \text{even} \ dx. \tag{13}$$

It is possible to justify Eqs. (12) and (13) by considering the areas under the given functions. Now, suppose $f(x)$ is an odd function on the interval $-a < x < a$.

Then by Eqs. (11) $f(x)\cos \dfrac{n\pi x}{2}$ is an odd function (since the cosine is an even function) and thus

$$a_n = \frac{1}{a} \int_{-a}^{a} f(x)\cos \frac{n\pi x}{a}\, dx = 0 \tag{14}$$

by Eq. (12) and

$$b_n = \frac{2}{a} \int_{0}^{a} f(x)\sin \frac{n\pi x}{a}\, dx \tag{15}$$

since $f(x)\sin \dfrac{n\pi x}{a}$ is an even function. Likewise

$$b_n = \frac{1}{a} \int_{-a}^{a} g(x)\sin \frac{n\pi x}{a}\, dx = 0 \tag{16}$$

and

$$a_n = \frac{2}{a} \int_{0}^{a} g(x)\cos \frac{n\pi x}{a}\, dx \tag{17}$$

whenever $g(x)$ is an even function on the interval $-a < x < a$.

Example 3.   Find the Fourier series representation for the function

$$f(x) = \begin{cases} -1 & -1 < x < 0 \\ 1 & 0 < x < 1 \end{cases}$$

and $f(x + 2) = f(x)$.

We note that $f(x)$ is an odd function of period 2 and thus $a = 1$ and Eqs. (14) and (15) yield $a_n = 0$ and

$$b_n = 2 \int_{0}^{1} 1 \cdot \sin n\pi x\, dx$$

$$= -\frac{2}{n\pi} \cos n\pi \Big|_0^1$$

$$= -\frac{2}{n\pi} [\cos n\pi - 1]$$

$$= \begin{cases} 0 & n \text{ even} \\ \dfrac{4}{n\pi} & n \text{ odd} \end{cases} .$$

Thus

$$f(x) = \frac{4}{\pi} \sum_{n=1}^{\infty} \frac{\sin(2n-1)\pi x}{(2n-1)} \tag{18}$$

is the desired Fourier series representation.

Other examples of the representation of even or odd functions were seen in Example 5. of the last section or Example 1. of this section.

The above examples and discussion tell us how to simplify the calculations of the a_n and the b_n if we know that a given function is either even or odd. However, if we know a given function which is neither even or odd on an interval, how do we get a Fourier series involving only cosines (or only sines)? To this end, let us consider a function $g(x)$ defined on the interval $0 < x < a$. Now, if it is desired to find a cosine series for this function we must extend $g(x)$ onto the interval $-a < x < 0$ so that $b_n = 0$. Thus we must extend $g(x)$ evenly onto the interval $-a < x < 0$, thereby obtaining an even function $f(x)$ defined for $-a < x < a$:

$$f(x) = \begin{cases} g(x) & 0 < x < a \\ g(-x) & -a < x < 0, \end{cases} \tag{19}$$

which is illustrated in Figure 2. Likewise, if we extend $g(x)$, defined for $0 < x < a$, as an odd function we obtain

$$h(x) = \begin{cases} g(x) & 0 < x < a \\ -g(-x) & -a < x < 0, \end{cases} \tag{20}$$

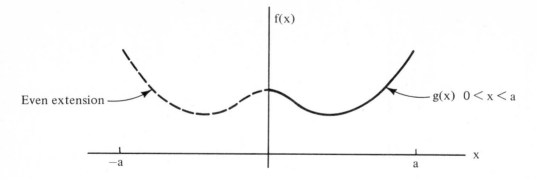

Figure 2.

which is illustrated in Figure 3.   In this case of course the a_n will be zero and
the Fourier representation will be in terms of sines only.

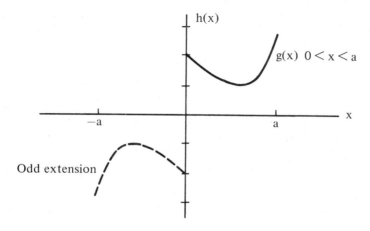

Figure 3.

Example 4.   Find the Fourier sine series and Fourier cosine series
                         representation of the function $g(x) = x$, $0 < x < 1$.

    For the sine series representation we have

$$h(x) = \begin{cases} x & 0 < x < 2 \\ -(-x) & -2 < x < 0 \end{cases} \tag{21}$$

    and thus

$$a_n = \frac{1}{2} \int_{-2}^{2} h(x)\cos \frac{n\pi x}{2}\, dx = 0$$

    and

$$b_n = \int_0^2 h(x)\sin\frac{n\pi x}{2}\, dx$$

$$= \int_0^2 x\sin\frac{n\pi x}{2}\, dx,$$

where Eqs. (14) and (15) have been used. Note that with Eq. (15) we only use values of the function on the original interval, and thus the extension given in Eq. (21) is purely formal. Evaluating the b_n we obtain

$$b_n = \frac{4}{\pi}\frac{(-1)^{n+1}}{n}$$

and thus

$$g(x) = \frac{4}{\pi}\sum_{n=0}^{\infty}\frac{(-1)^{n+1}\sin\frac{n\pi x}{2}}{n}\,. \qquad (22)$$

For the cosine series representation we have

$$f(x) = \begin{cases} x & 0 < x < 2 \\ (-x) & -2 < x < 0 \end{cases} \qquad (23)$$

and thus

$$a_n = \int_0^2 f(x)\cos\frac{n\pi x}{2}\, dx$$

$$= \int_0^2 x\cos\frac{n\pi x}{2}\, dx \qquad (24)$$

and

$$b_n = \frac{1}{2}\int_{-2}^2 f(x)\sin\frac{n\pi x}{2}\, dx = 0$$

where Eqs. (16) and (17) have been used. Note again that Eq. (23) is purely

formal.   The values of a_n in Eq. (24) are identical with those of Example 1.
and thus

$$g(x) = 1 - \frac{8}{\pi^2} \sum_{n=1}^{\infty} \frac{\cos\frac{(2n-1)\pi x}{2}}{(2n-1)^2} \tag{25}$$

is the desired cosine series representation.

It should be emphasized that if a function is defined on $0 < x < a$, then the
Fourier sine series or cosine series representation will converge to a function
which is periodic of period $2a$ since the even or odd extension involves the interval
$-a < x < a$.   However, the original function will still be represented for $0 < x < a$.
On the other hand, if $f(x)$ is defined on $0 < x < a$ and it is desired to have a representation of period   a, only, then generally both sine and cosine terms will be
involved since the periodic extension may be neither even or odd.

Exercises

1.   Classify the following functions as even, odd or neither.

a) $x^4 - 2x^2 + 1$ b) $e^x + e^{-x}$

c) $\sin(x + 2)$ d) $x^2 - 2 \cos 3x$

e) $\ln|x|$ f) $x^3 + x^5 - 2x$

2.   Find the Fourier series representation for each of the following functions.
     Sketch the graphs for at least two periods.

a) $f(x) = \begin{cases} -1 & -2 < x < 0 \\ x & 0 < x < 2 \end{cases}$, $f(x + 4) = f(x)$

b) $f(x) = x^2$ $-1 < x < 1$, $f(x + 2) = f(x)$

c) $f(x) = \begin{cases} 3 - x & 0 < x < 3 \\ -3 - x & -3 < x < 0 \end{cases}$, $f(x + 6) = f(x)$

d) $f(x) = e^x$ $-1 < x < 1$, $f(x + 2) = f(x)$

e) $f(x) = 1$ $-2 < x < 2$, $f(x + 4) = f(x)$

f) $f(x) = \sin \pi x$ $-1 < x < 1$, $f(x + 2) = f(x)$

3. Show that any function $f(x)$ can be written as the sum of an even and an odd function. (Hint: Assume $f(x) = g(x) + h(x)$, where $g(x)$ is even and $h(x)$ is odd, and find $f(-x)$.)

4. Justify the relationships of Eq. (11).

5. If $f(x)$ is an odd function show that $\displaystyle\int_{-a}^{a} f(x)dx = 0$.

6. If $f(x)$ is an even function show that $\displaystyle\int_{-a}^{a} f(x)dx = 2 \int_{0}^{a} f(x)dx$.

7. Find the Fourier cosine and sine series of period 2 for the function $f(x) = 1 + x$, $0 < x < 1$. Sketch the graphs for at least two periods.

8. Find the Fourier cosine and sine series of period 2ℓ for the function $f(x) = \ell - x$, $0 < x < \ell$. Sketch the graphs for at least 2 periods.

9. Find the Fourier series of period 1 for the function $f(x) = x$, $0 < x < 1$.

10. Suppose that $f(x)$ is continuous for $0 < x < a$. Is its even periodic extension of period $2a$ also continuous? What about the odd periodic extension of period $2a$?

11. Show that the derivative of an even function is odd and that the derivative of an odd function is even.

12. If $F(x) = \displaystyle\int_{0}^{x} f(t)dt$,   show that if $f(x)$ is even then $F(x)$ is odd and if $f(x)$ is odd then $F(x)$ is even.

13. Show that the functions $\cos \dfrac{n\pi x}{a}$ and $\sin \dfrac{n\pi x}{a}$ satisfy orthogonality relations on the interval $-a < x < a$ similar to those of Section 8.1.

14. Show that the orthogonality relations

$$\int_0^a \sin\frac{n\pi x}{a} \sin\frac{m\pi x}{a}\,dx \;=\; \begin{cases} 0 & n \neq m \\[2mm] a/2 & n = m \end{cases}$$

and

$$\int_0^a \cos\frac{m\pi x}{a} \cos\frac{n\pi x}{a}\,dx \;=\; \begin{cases} 0 & n \neq m \\[2mm] a/2 & n = m \neq 0 \\[2mm] a & n = m = 0 \end{cases}$$

follow from those of Exercise 13.

8.3   CONVERGENCE PROPERTIES AND APPLICATIONS OF FOURIER SERIES

In the last two sections we have derived the necessary formulas and concepts for obtaining the Fourier series representation of a large variety of periodic functions. For reference here, what we have found is that if $f(x)$ is periodic of period $2a$, then its Fourier series representation is given by

$$f(x) \;=\; \frac{a_0}{2} + \sum_{n=1}^{\infty} a_n \cos\frac{n\pi x}{a} + b_n \sin\frac{n\pi x}{a} \tag{1}$$

when

$$a_n \;=\; \frac{1}{a} \int_{-a}^{a} f(x)\cos\frac{n\pi x}{a}\,dx \tag{2}$$

and

$$b_n \;=\; \frac{1}{a} \int_{-a}^{a} f(x)\sin\frac{n\pi x}{a}\,dx \; . \tag{3}$$

The question that needs to be answered now is: if a value of x is substituted into the series in Eq. (1) then does the resulting series add up to the functional value of $f(x)$ when the a_n and b_n are given by Eqs. (2) and (3)? If the answer to this question is yes, then the use of the equality in Eq. (1) has been justified. In this section we shall present some results which answer the question, but first we must review some concepts of limits and continuity.

In looking over the examples and exercises of the previous two sections, we see that most of the functions for which we have calculated the Fourier series representation exhibit what is known as <u>jump</u> discontinuities (these were also encountered in Section 7.4). A jump discontinuity occurs at a point x_0 provided the <u>limits</u> as

$x \to x_0$ from the right and from the left both exist but are not equal.

Example 1.   The function

$$f(x) = \begin{cases} 1 - x & 0 < x < 1 \\ -1 + x & -1 < x < 0 \end{cases}$$

as shown in Figure 1. has a jump discontinuity at $x = 0$.   This is

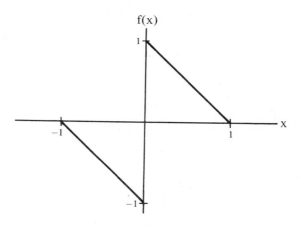

Figure 1.

justified analytically since $\lim\limits_{x \to 0^+} f(x) = 1$ and $\lim\limits_{x \to 0^-} f(x) = -1$,

where $x \to 0^+$ means x approaches 0 from the right and $x \to 0^-$ means
x approaches 0 from the left. Thus the two limits exist, but are
not equal.

A second special type of discontinuity that is of concern here is that of a
removable discontinuity (these were also encountered in Section 6.2). In this case
the right and left hand limits both exist and are equal, but the function is not
defined at the point.

Example 2.   The function

$$f(x) = \begin{cases} 1 - x & 0 < x < 1 \\ 1 + x & -1 < x < 0 \end{cases}$$

as shown in Figure 2. has a removable discontinuity at $x = 0$ since
$\lim\limits_{x \to 0^+} f(x) = \lim\limits_{x \to 0^-} f(x) = 1$ but $f(0)$ has not been defined.

The type of discontinuity exhibited in Example 2. is called removable since it is

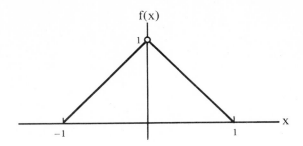

Figure 2.

always possible to find a new function which agrees with f(x), where it is defined, and yet is continuous at points where f(x) has removable discontinuities. For instance if f(x) is as defined in Example 2, then the function g, given by g(x) = f(x) for x ≠ 0, and g(0) = 1, is continuous for -1 < x < 1.

Other types of discontinuities are exhibited by the functions sin(1/x) and 1/x at x = 0. In these cases the right and left hand limits don't even exist and thus are of no interest here since in general their Fourier series representation does not exist.

Functions which are continuous on a finite interval except for a finite number of jump or removable discontinuities are known as <u>piecewise</u> <u>continuous</u> functions. Furthermore, if f'(x) is piecewise continuous as well as f(x), then f(x) is said to be <u>piecewise</u> <u>smooth</u>. All of the example functions in this and the last two sections have been piecewise smooth. The fact that the functions have been piecewise continuous can be seen by inspection. If each of them is differentiated, then it can be seen that the derivatives are also piecewise continuous. It should be pointed out that the derivative may not exist at a point, but this is acceptable as long as the left and right hand limits exist.

Example 3.   The function $f(x) = |x|$ is continuous for all x   and

$$f'(x) = \begin{cases} 1 & 0 < x \\ -1 & x < 0 \,. \end{cases}$$

Thus f(x) is said to be piecewise smooth since f'(x) is piecewise continuous.

Example 4.   The function $f(x) = |x|^{\frac{1}{2}}$ is continuous for all x, and therefore it is piecewise continuous. However it is not piecewise smooth since f'(x) is unbounded as x approches zero.

Functions which are continuous on a finite interval (say $-a < x < a$) and have a derivative which is piecewise continuous on the same interval will have periodic extensions that are piecewise smooth.

Example 5.   The function

$$f(x) = x, \qquad\qquad -1 < x < 1$$

is continuous and has a continuous derivative ($f'(x) = 1$, $-1 < x < 1$) on the interval of definition. Its periodic extension, the saw tooth wave, is therefore piecewise smooth.

We are now in a position to state the convergence results that will justify the use of equality in our previous work. Let $f(x^+)$ and $f(x^-)$ denote the right and left hand limits of $f(x)$ at x. Then if $f(x)$ is piecewise smooth and periodic with period $2a$, then at each point x we have

$$\frac{f(x^+) + f(x^-)}{2} = \frac{a_0}{2} + \sum_{n=1}^{\infty} a_n \cos \frac{n\pi x}{a} + b_n \sin \frac{n\pi x}{a}, \qquad (4)$$

where the a_n and b_n are given by Eqs. (2) and (3). This is known as pointwise convergence since for each x the series adds up to the left hand side of Eq. (4). If $f(x)$ is continuous at a point x, then the left hand side of Eq. (4) reduces to $f(x)$ since $f(x^+) = f(x^-)$ in this case. Otherwise the left hand side is the average value of $f(x)$ at a jump discontinuity. Since there are only a finite number of these for a piecewise smooth function, we simply write f equal to its Fourier series and interpret this to mean the average value at the jump discontinuities.

Example 6.   From Example 5. of Section 8.1 we saw that the saw-tooth wave of Figure 3. has the Fourier series

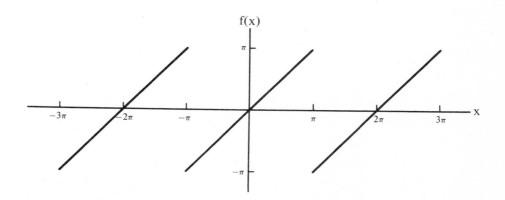

Figure 3.

$$f(x) = \sum_{n=1}^{\infty} \frac{2(-1)^{n+1}}{n} \sin nx. \tag{5}$$

From the graph we see that $f(x)$ is piecewise smooth and has jump discontinuities at $x = \pi, 3\pi, \cdots$, where the average values of the left $(+1)$ and right (-1) hand limits is always zero. By substituting $x = m\pi$ into Eq. (5) we see indeed that the series does add up to zero at each of these points. For $x = \pi/2$, we can apply the results of Eq. (4) to obtain

$$\frac{\pi}{2} = \sum_{n=1}^{\infty} \frac{2(-1)^{n+1}}{n} \sin \frac{n\pi}{2}$$

$$= \sum_{k=1}^{\infty} \frac{2}{2k-1} \sin \frac{(2k-1)\pi x}{2},$$

since $\sin \frac{n\pi}{2} = 0$ for even n. Finally, since $\sin \frac{(2k-1)\pi}{2}$ is either $+1$ or -1 for integer k, we obtain

$$\frac{\pi}{4} = \sum_{k=1}^{\infty} \frac{(-1)^{k+1}}{(2k-1)} \tag{6}$$

$$= 1 - \frac{1}{3} + \frac{1}{5} - \frac{1}{7} + \cdots .$$

There are other types of convergence properties that could be discussed, such as mean convergence. However, for most applications, the pointwise convergence for piecewise smooth functions is sufficient. In applications it is often necessary to differentiate or integrate a Fourier series representation of a function. If $f(x)$ is piecewise continuous then it is always possible to integrate termwise. However, in order to differentiate a Fourier series termwise, it is necessary that $f'(x)$ be piecewise smooth and $f(x)$ be continuous.

Example 7.   From Example 4. of Section 8.2 we found that

$$f(x) = 1 - \frac{8}{\pi^2} \sum_{m=1}^{\infty} \frac{\cos \frac{(2m-1)\pi x}{2}}{(2m-1)^2}, \tag{7}$$

is the Fourier series for $f(x) = |x|$, $-2 < x < 2$ and $f(x + 4) = f(x)$.
Thus

$$f'(x) = \frac{4}{\pi} \sum_{m=1}^{\infty} \frac{\sin \frac{(2m-1)\pi x}{2}}{2m - 1} \tag{8}$$

is the termwise differentiation of Eq. (7). Since $f(x) = |x|$
$-2 < x < 2$, its derivative is given by

$$f'(x) = \begin{cases} 1 & 0 < x < 2 \\ -1 & -2 < x < 0 \end{cases}$$

so that $f'(x)$ is piecewise smooth and $f(x)$ is continuous. Therefore
the series in Eq. (8) does indeed converge to $f'(x)$.

As mentioned previously, Fourier series representation of functions occur very
frequently in a number of applications. Two of these that are easily demonstrated
will be presented here. The first concerns the solution of the forced vibration
problem

$$\frac{d^2 y}{dx^2} + \lambda^2 y = f(x), \tag{9}$$

$$y(0) = 0, \quad y(a) = 0. \tag{10}$$

These problems were briefly introduced in Section 1.3 where for certain functions
$f(x)$ the general solution of Eq. (9) could be found. For more general functions,
or for a different representation of the solution $y(x)$ we are led to consider representing the solution as a Fourier series on the interval $0 < x < a$. This is reason-
able since $f(x)$ is given on that interval and in general, for this type problem,
$y(x)$ is desired only for the given interval. Since Eqs. (10) must also be satisfied,
we are led to consider a sine series representation of $y(x)$:

$$y(x) = \sum_{n=1}^{\infty} b_n \sin \frac{n\pi x}{a} \tag{11}$$

as this automatically satisfies both boundary conditions in Eqs. (10). In this case,
however, the coefficients b_n cannot be found using Eq. (3) as the $y(x)$ is not known.
However, if $f(x)$ is assumed piecewise smooth on the interval $0 < x < a$, then Eq. (9)
implies that $y(x)$ and $y'(x)$ are continuous and $y''(x)$ is piecewise smooth. Hence
Eq. (11) can be differentiated twice to give

$$y''(x) = - \sum_{n=1}^{\infty} (\frac{n^2 \pi^2}{a^2} b_n) \sin \frac{n\pi x}{a} . \tag{12}$$

Finally, $f(x)$ has the Fourier sine series representation:

$$f(x) = \sum_{n=1}^{\infty} c_n \sin \frac{n\pi x}{a} \tag{13}$$

where

$$c_n = \frac{2}{a} \int_0^a f(x) \sin \frac{n\pi x}{a} dx. \tag{14}$$

If Eqs. (11), (12) and (13) are now substituted into Eq. (9) we obtain

$$\sum_{n=1}^{\infty} (\lambda^2 - \frac{n^2 \pi^2}{a^2}) b_n \sin \frac{n\pi x}{a} = \sum_{n=1}^{\infty} c_n \sin \frac{n\pi x}{a} \tag{15}$$

after the terms on the left are combined. Since the sine terms are the same on both sides, Eq. (15) is satisfied only if corresponding coefficients are equal. Thus the b_n are given by

$$b_n = \frac{c_n}{\lambda^2 - (n\pi/a)^2} \tag{16}$$

provided λ is not a multiple of π/a. If λ is a multiple of π/a, then Eqs. (9) and (10) have no solution unless the corresponding c_n is zero (in this case there is an arbitrary constant in the solution). Summarizing, then, the solution to Eqs. (9) and (10) is given by

$$y(x) = \sum_{n=1}^{\infty} \frac{c_n \sin \frac{n\pi x}{a}}{\lambda^2 - (n\pi/a)^2} \tag{17}$$

where c_n is given by Eq. (14) and λ is not a multiple of π/a.

The second application of Fourier series concerns the solution of constant coefficient second order partial differential equations. There are basically three types of equations that can be solved in this manner, but the techniques are similar, so only one will be presented here. If $T(x)$ represents the temperature in a pipe ℓ units long at $t = 0$, then the temperature $u(x,t)$ at a later time is given by

$$\frac{\partial^2 u}{\partial x^2} = \frac{1}{k} \frac{\partial u}{\partial t}, \qquad\qquad 0 < x < \ell, \; t > 0 \qquad\qquad (18)$$

$$\frac{\partial u}{\partial x}(0,t) = 0, \; \frac{\partial u}{\partial x}(\ell,t) = 0, \qquad\qquad t > 0 \qquad\qquad (19)$$

$$u(x,0) = T(x), \qquad\qquad 0 < x < \ell, \qquad\qquad (20)$$

if the pipe is entirely insulated. The constant k is determined by the material properties of the pipe. To solve the problem posed by Eqs. (18), (19) and (20) we note that $\cos \frac{n\pi x}{\ell}$ satisfies the boundary conditions (19) for any integer n. Thus we are led to consider

$$u(x,t) = \frac{a_0(t)}{2} + \sum_{n=1}^{\infty} a_n(t) \cos \frac{n\pi x}{\ell}, \qquad\qquad (21)$$

where the coefficients can be functions of t since u is also a function of t. The $u(x,t)$ given in Eq. (21) satisfies Eqs. (19) and it is now necessary to find $a_n(t)$ so that it satisfies Eq. (18). Using the approach shown for the previous application, we can differentiate Eq. (21) ($u(x,t)$ can be shown to have the required continuity properties), substitute it into Eq. (18) and equate like coefficients to yield

$$a_0'(t) = 0$$

$$(22)$$

$$a_n'(t) = -\beta_n a_n'(t), \; n = 1,2, \cdots,$$

where $\beta_n = n^2\pi^2 k/\ell^2$. The solution of Eqs. (22) are

$$a_n(t) = c_n e^{-\beta_n t}, \qquad n = 0,1,2, \cdots$$

and thus $u(x,t)$ becomes

$$u(x,t) = \frac{c_0}{2} + \sum_{n=1}^{\infty} c_n e^{-\beta_n t} \cos \frac{n\pi x}{\ell}, \qquad\qquad (23)$$

which now satisfies Eqs. (18) and (19). Letting $t = 0$ in Eq. (23) we obtain

$$u(x,0) = \frac{c_0}{2} + \sum_{n=1}^{\infty} c_n \cos \frac{n\pi x}{\ell}, \qquad\qquad (24)$$

which by Eq. (20) must equal $T(x)$. Since $T(x)$ is given on the interval $0 < x < \ell$, we can calculate the c_n in Eq. (24) by considering the even periodic extension of $T(x)$. Thus if

$$c_n = \frac{2}{\ell} \int_0^\ell T(x) \cos \frac{n\pi x}{\ell} dx, \tag{25}$$

then Eq. (24) will add up to T(x) (assuming T(x) is piecewise smooth on $0 < x < \ell$). Summarizing, then, we have shown that Eq. (23) when the coefficients c_n are given by Eq. (25), is the solution to the problem posed by Eqs. (18), (19) and (20).

Exercises

1. Determine if each of the following functions is piecewise smooth. If it is, state the value to which its Fourier series converges at each point x, including the end points. Sketch each function for at least two periods.

   a) $f(x) = |x|$ $-1 < x < 1,$ $f(x + 2) = f(x)$

   b) $f(x) = x + |x|$ $-1 < x < 1,$ $f(x + 2) = f(x)$

   c) $f(x) = x \sin x$ $-1 < x < 1,$ $f(x + 2) = f(x)$

   d) $f(x) = x \sin x$ $-\pi < x < \pi,$ $f(x + 2\pi) = f(x)$

   e) $f(x) = \begin{cases} 0 & -2 < x < -1 \\ 1 + x & -1 < x < 1 \\ 1 & 1 < x < 2 \end{cases}$ $f(x + 4) = f(x)$

2. Suppose $f(x) = 1 - x^2/4$, $0 \le x \le 1$. To what value does the Fourier cosine series of period 2 converge to at $x = 0$, 1 and -1? To what value does the Fourier sine series of period 2 converge to at $x = 0$, 1, and -1?

3. Given the Fourier series

$$|x| = \frac{1}{2} - \frac{4}{\pi^2} \sum_{n=1}^{\infty} \frac{\cos(2n-1)\pi x}{(2n - 1)^2}, \quad -1 < x < 1$$

   show that

$$\frac{\pi^2}{8} = 1 + \frac{1}{9} + \frac{1}{25} + \cdots$$

by choosing an appropriate value of x and applying the pointwise convergence result.

4. Repeat Exercise 3. for

$$\left| \sin x \right| = \frac{1}{\pi} - \frac{2}{\pi} \sum_{n=1}^{\infty} \frac{\cos 2nx}{4n^2 - 1}$$

and

$$\frac{1}{2} = \frac{1}{3} + \frac{1}{15} + \frac{1}{35} + \cdots .$$

5. Use the Fourier series approach developed in this section to solve each of the following problems.

a) $y'' + 2y = x,$ $\qquad$ $y(0) = 0,$ $\quad y(3) = 0$

b) $y'' + \frac{1}{2}y = e^x,$ $\qquad$ $y(0) = 0,$ $\quad y(\pi) = 0$

c) $y'' + 3y = 2\sin 3\pi x,$ $\qquad y(0) = 0,$ $\quad y(1) = 0$

6. If the forcing function $f(t)$ is periodic of period $2a$ then a particular solution of the differential equation

$$\frac{d^2 y}{dt^2} + \alpha \frac{dy}{dt} + \beta y = f(t)$$

can be found using Fourier series. Letting

$$f(t) = \frac{a_0}{2} + \sum_{n=1}^{\infty} a_n \cos \frac{n\pi t}{a} + b_n \sin \frac{n\pi t}{a}$$

and

$$y(t) = \frac{A_0}{2} + \sum_{n=1}^{\infty} A_n \cos \frac{n\pi t}{a} + B_n \sin \frac{n\pi t}{a},$$

find a formula for the A_n and B_n in terms of the a_n and b_n.

7. If $f(t)$ is given by

$$f(t) = \begin{cases} 1 & 0 < t < 1 \\ 0 & 1 < t < 2 \end{cases}, \qquad f(t + 2) = f(t)$$

find a particular solution of $\dfrac{d^2 y}{dt^2} + \alpha \dfrac{dy}{dt} + \beta y = f(t).$

8.   Find the solution to Eqs. (18), (19) and (20) for each of the following initial
     temperatures.

     a) $T(x)$ $=$ $\begin{cases} 1 & 0 < x < \ell/2 \\ 2 & \ell/2 < x < \ell \end{cases}$

     b) $T(x)$ $=$ x $0 < x < \ell$

9.   Show that the solution to Eqs. (18), (19) and (20) has the property that
     $\lim\limits_{t\to\infty} u(x,t) = \dfrac{a_0}{2}$.   Thus the final (steady state) temperature in the

     pipe is equal to the average value of the original temperature, independent
     of the original temperature distribution.

10. Find the form of the solution to

     $$\frac{\partial^2 u}{\partial x^2} = \frac{1}{k} \frac{\partial u}{\partial t} \qquad\qquad 0 < x < \ell, \qquad t > 0$$

     $$u(0,t) = 0, \quad u(\ell,t) = 0, \qquad t > 0$$

     $$u(x,0) = T(x), \qquad\qquad 0 < x < \ell.$$

11. Find the form of the solution to

     $$\frac{\partial^2 u}{\partial x^2} = \frac{1}{c^2} \frac{\partial^2 u}{\partial t^2} \qquad\qquad 0 < x < \ell, \qquad t > 0$$

     $$u(0,t) = 0, \quad u(\ell,t) = 0, \qquad t > 0$$

     $$u(x,0) = f(x) \qquad\qquad 0 < x < \ell$$

     $$\frac{\partial u}{\partial t}(x,0) = g(x) \qquad\qquad 0 < x < \ell,$$

     assuming $u(x,t)$ can be represented as in Eq. (21).

8.4   FOURIER INTEGRALS

In the preceding sections of this chapter we have shown how to represent functions that are periodic of period 2a, where for the most part 2a is not very large. A very natural extension of those considerations is what occurs to the representation when 2a is very large, and in particular, when 2a approaches infinity. In the application problem of the preceding section, which considered the temperature distribution in a pipe, the length of many pipes is such that for all practical purposes the pipe can be considered "infinitely" long. In this section, then, we will give a very brief introduction to the concepts of Fourier integral representation of functions, defined for all x but which need not be periodic.

The Fourier integral representation can be considered as long as f(x) is piecewise smooth on every finite interval (which follows from the above discussion) and assumes that $\int_{-\infty}^{\infty} |f(x)|dx$ is finite. Then it can be shown for every point x   that

$$\frac{f(x^+) + f(x^-)}{2} = \int_0^{\infty} \{A(\lambda)\cos \lambda x + B(\lambda)\sin \lambda x\}d\lambda \tag{1}$$

when

$$A(\lambda) = \frac{1}{\pi} \int_{-\infty}^{\infty} f(x)\cos(\lambda x)dx \tag{2}$$

and

$$B(\lambda) = \frac{1}{\pi} \int_{-\infty}^{\infty} f(x)\sin(\lambda x)dx. \tag{3}$$

Notice that the form is similar to the Fourier series encountered previously, where the infinite sum has become the integral of Eq. (1) and at jump discontinuities the average value is the appropriate representation.

Example 1.   Find the Fourier integral representation of

$$f(x) = \begin{cases} e^{-x} & 0 < x \\ -e^{x} & x < 0 . \end{cases} \tag{4}$$

Since f(x) is piecewise smooth and since $\int_{-\infty}^{\infty} |f(x)|dx = 2\int_0^{\infty} e^{-x}\, dx$

is finite we can use Eqs. (2) and (3). Since f(x) is an odd function
we obtain

$$A(\lambda) = \frac{1}{\pi} \int_{-\infty}^{\infty} f(x)\cos(\lambda x)dx = 0$$

and

$$B(\lambda) = \frac{1}{\pi} \int_{-\infty}^{\infty} f(x)\sin(\lambda x)dx$$

$$= \frac{2}{\pi} \int_{0}^{\infty} e^{-x} \sin(\lambda x)dx$$

$$= -\frac{2}{\pi} \frac{(\sin \lambda x + \lambda \cos \lambda x)e^{-x}}{1 + \lambda^2} \Bigg|_{0}^{\infty}$$

$$= \frac{2}{\pi} \frac{\lambda}{1 + \lambda^2} \ . \tag{5}$$

Thus Eq. (1) yields

$$f(x) = \frac{2}{\pi} \int_{0}^{\infty} \frac{\lambda \sin \lambda x}{1 + \lambda^2} d\lambda, \tag{6}$$

where, as previously denoted, the value of f(x) on the left hand side of
Eq. (6) is interpreted as the average value of the right and left hand
limits at points where f(x) has jump discontinuities. In this case f(x)
has a jump discontinuity at x = 0, the average value being 0, which is
exactly what the right hand side of Eq. (6) yields when x = 0.

The Fourier integral representation can be formally obtained from the Fourier
series representation, although the limiting process cannot be justified. However,
since essentially correct results are obtained it will be outlined here to illustrate
the process and to show the relationships between the two representations. If f(x)
is piecewise smooth in every finite interval, then in the interval -a < x < a

$$f(x) = \frac{a_0}{2} + \sum_{n=1}^{\infty} a_n \cos \frac{n\pi x}{a} + b_n \sin \frac{n\pi x}{a} \tag{7}$$

when

$$a_n = \frac{1}{a} \int_{-a}^{a} f(x)\cos\frac{n\pi x}{a}\, dx \tag{8}$$

and

$$b_n = \frac{1}{a} \int_{-a}^{a} f(x)\sin\frac{n\pi x}{a}\, dx . \tag{9}$$

Comparing Eqs. (8) and (9) with Eqs. (2) and (3) we see that if we define

$$\lambda_n = \frac{n\pi}{a}, \quad A(\lambda_n) = \frac{a}{\pi} a_n \quad \text{and} \quad B(\lambda_n) = \frac{a}{\pi} b_n \quad \text{we obtain}$$

$$f(x) = \frac{a_0}{2} + \sum_{n=1}^{\infty} \{A(\lambda_n)\cos\lambda_n x + B(\lambda_n)\sin\lambda_n x\}\Delta\lambda \tag{10}$$

where

$$A(\lambda_n) = \frac{1}{\pi} \int_{-a}^{a} f(x)\cos(\lambda_n x)dx, \tag{11}$$

$$B(\lambda_n) = \frac{1}{\pi} \int_{-a}^{a} f(x)\sin(\lambda_n x)dx, \tag{12}$$

and

$$\Delta\lambda = \pi/a.$$

Now, as $a \to \infty$, $\Delta\lambda \to 0$ and the infinite sum in Eq. (10) suggests an integral. If

$$\int_{-\infty}^{\infty} |f(x)|dx \quad \text{is finite, then } a_0 = \frac{1}{a} \int_{-a}^{a} f(x)dx \quad \text{approaches zero as } a \to \infty \text{ and thus}$$

Eqs. (10), (11) and (12) become Eqs. (1), (2) and (3) as $a \to \infty$.

For functions which are only defined for $0 < x$, we can still apply Eqs. (2) and (3) to find their integral representation by extending the function to the whole range of x values. If even and odd periodic extensions are used, then we are led to the following Fourier cosine and Fourier sine integral representations respectively. If $f(x)$ is extended evenly we obtain

$$f(x) = \int_{0}^{\infty} A(\lambda)\cos(\lambda x)d\lambda \qquad 0 < x \tag{13}$$

where

$$A(\lambda) = \frac{2}{\pi} \int_0^\infty f(x)\cos(\lambda x)dx \tag{14}$$

and if $f(x)$ is extended oddly we obtain

$$f(x) = \int_0^\infty B(\lambda)\sin(\lambda x)d\lambda \qquad 0 < x \tag{15}$$

where

$$B(\lambda) = \frac{2}{\pi} \int_0^\infty f(x)\sin(\lambda x)d\lambda. \tag{16}$$

Notice, as with Fourier series, that the extensions are purely formal, as only information about $f(x)$ is needed on $0 < x$. Referring back to Example 1, we see that what was obtained there was a Fourier sine integral representation since the given function was already defined as an odd function.

Example 2.   Find the Fourier cosine integral representation of
$f(x) = e^{-x}$, $0 < x$.

Extending $f(x)$ evenly throughout $-\infty < x < \infty$, we can use Eq. (14) to find $A(\lambda)$:

$$A(\lambda) = \frac{2}{\pi} \int_0^\infty e^{-x}\cos(\lambda x)dx$$

$$= \frac{2}{\pi} \left. \frac{(\lambda \sin \lambda x - \cos \lambda x)e^{-x}}{1 + \lambda^2} \right|_0^\infty$$

$$= \frac{2}{\pi} \frac{1}{1 + \lambda^2}$$

and therefore

$$e^{-x} = \frac{2}{\pi} \int_0^\infty \frac{\cos \lambda x}{1 + \lambda^2} d\lambda, \qquad 0 < x. \tag{17}$$

Of course, it is possible, to extend functions in other ways. In these cases, however, the complete integral representation will be obtained. In some applications these alternate representations may be needed.

Example 3.   Find the Fourier integral representation of $f(x) = e^{-x}$,
                    $0 < x$, by extending it to be zero for $x < 0$.

Letting

$$g(x) = \begin{cases} e^{-x} & 0 < x \\ 0 & x < 0 \end{cases}$$

we see that

$$A(\lambda) = \frac{1}{\pi} \int_{-\infty}^{\infty} g(x)\cos(\lambda x)dx$$

$$= \frac{1}{\pi} \int_{0}^{\infty} e^{-x} \cos(\lambda x)dx = \frac{1}{\pi} \frac{1}{1 + \lambda^2}$$

and that

$$B(\lambda) = \frac{1}{\pi} \int_{-\infty}^{\infty} g(x)\sin(\lambda x)dx$$

$$= \frac{1}{\pi} \int_{0}^{\infty} e^{-x} \sin(\lambda x)dx = \frac{1}{\pi} \frac{\lambda}{1 + \lambda^2}$$

since $g(x) = 0$ for $x < 0$.   Thus

$$e^{-x} = \frac{1}{\pi} \int_{0}^{\infty} \frac{\cos \lambda x + \lambda \sin \lambda x}{1 + \lambda^2} d\lambda \tag{18}$$

is a Fourier integral representation of e^{-x} for $0 < x$.   For $x < 0$
Eq. (1) would yield

$$0 = \frac{1}{\pi} \int_{0}^{\infty} \frac{\cos \lambda x + \lambda \sin \lambda x}{1 + \lambda^2} d\lambda \tag{19}$$

and for $x = 0$ we obtain

$$\frac{1}{2} = \frac{1}{\pi} \int_{0}^{\infty} \frac{d\lambda}{1 + \lambda^2} \tag{20}$$

since $\dfrac{g(0^+) + g(0^-)}{2}$ is $\dfrac{1}{2}$.

As an application of the Fourier integral consider the problem of determining the temperature in an "infinitely" long pipe (if ℓ, of the last section, is very large compared to the radius the "infinitely" long concept is appropriate). In this case let the end at x = 0 be held at 0 temperature and let the original temperature at each point x be dentoed by T(x). Then the temperature at any later time is given by

$$\frac{\partial^2 u}{\partial x^2} = \frac{1}{k}\frac{\partial u}{\partial t}, \qquad\qquad x > 0 \text{ and } t > 0 \qquad\qquad (21)$$

$$u(0,t) = 0, \qquad\qquad t > 0 \qquad\qquad (22)$$

$$u(x,0) = T(x), \qquad\qquad x > 0. \qquad\qquad (23)$$

Using our experience with this problem in the last section we see that it is appropriate to consider u(x,t) written in the form

$$u(x,t) = \int_0^\infty B(\lambda)\sin(\lambda x)e^{-\lambda^2 kt}\, d\lambda. \qquad\qquad (24)$$

Assuming that differentiation under the integral is appropriate, it can be verified that u(x,t) as given in Eq. (24) satisfies Eq. (21) and by setting x = 0, Eq. (22) can be seen to be satisfied. Thus, in order to solve the complete problem, we must find B(λ) so that Eq. (23) is satisfied. Setting t = 0 in Eq. (24) we see that B(λ) must be chosen so that

$$T(x) = \int_0^\infty B(\lambda)\sin(\lambda x)d\lambda. \qquad\qquad (25)$$

Hence, if T(x) is piecewise smooth and $\int_0^\infty |T(x)|dx$ is finite, then Eq. (25) says B(λ) must be given by

$$B(\lambda) = \frac{2}{\pi}\int_0^\infty T(x)\sin(\lambda x)dx. \qquad\qquad (26)$$

Therefore Eqs. (24) and (26) give the solution to the problem posed by Eqs. (21), (22) and (23).

The concepts of Fourier series, Fourier integrals and the Laplace transforms (Chapter 7.) are all related to the general integral transform of a function f(t), which is defined as

$$F(s) = \int_a^b K(s,t)f(t)dt, \tag{27}$$

where $K(s,t)$ is called a __kernel__. In particular if $K(s,t) = e^{-st}$ with $a = 0$, $b = \infty$ we obtain

$$F(s) = \int_0^\infty e^{-st} f(t)dt \tag{28}$$

which is the Laplace transform of $f(t)$. In the light of the representations of this chapter, we see that the inverse Laplace transform

$$f(t) = \frac{1}{2\pi i} \int_{c-i\infty}^{c+i\infty} e^{st} F(s)ds \tag{29}$$

of Section 7.5 can be interpreted as an integral representation of $f(t)$. Likewise the Fourier sine transform is obtained when $K(s,t) = \frac{2}{\pi} \sin(st)$ and $a = 0$, $b = \infty$:

$$F(s) = \frac{2}{\pi} \int_0^\infty f(t)\sin(st)dt, \tag{30}$$

which is the same as $B(\lambda)$ of Eq. (16) if s is replaced by λ. Again, the inverse transform of Eq. (30) yields the integral representation of $f(t)$ found in Eq. (15). In a similar manner Eqs. (2), (3) and (14) can be obtained from Eq. (27) for different choices of $K(s,t)$, and are left as exercises.

To obtain the Fourier series work from Eq. (27) we must consider s a discrete variable ($s = n$) and then pick $K(n,t) = \frac{2}{\ell} \cos \frac{n\pi x}{\ell}$ with $a = 0$ and $b = \ell$. Thus Eq. (27) becomes

$$F(n) = \frac{2}{\ell} \int_0^\ell f(t) \cos \frac{n\pi t}{\ell} dt, \tag{31}$$

which is the same as a_n for the Fourier cosine series. As above then, the series representation is obtained from Eq. (31) by finding the inverse transform, which will be a series instead of an integral because n in Eq. (31) is discrete rather than a continuous variable.

Two other well known transforms are obtained by setting $K(s,t) = t^{s-1}$, which leads to the Mellin transform:

$$F(s) = \int_0^\infty f(t) t^{s-1} dt \tag{32}$$

and setting $K(s,t) = \dfrac{1}{\pi(t-s)}$, which leads to the Hilbert transform

$$F(s) = \frac{1}{\pi} \int_0^\infty \frac{f(t)}{t-s} dt. \tag{33}$$

The inverse transforms of these equations lead to two alternate representations of functions $f(t)$.

This has been a brief introduction to the concepts of Fourier series and integrals. However, the main points have been covered and examples have been worked so that their uses can be understood. As Eq. (27) has pointed out, there are many ways that this material can be extended in more advanced work.

Exercises

1. Find the Fourier integral representation of each of the following functions.

a) $f(x) = \begin{cases} 0 & x < 0 \\ 1/2 & 0 < x < a \\ 0 & a < x \end{cases}$ b) $f(x) = e^{-|x|}$

c) $f(x) = \dfrac{x}{1 + x^2}$ d) $f(x) = e^{-x^2/2a^2}$

2. Find the Fourier sine and cosine integrals for each of the following functions.

a) $f(x) = \begin{cases} 1/a & 0 < x < a \\ 0 & a < x \end{cases}$ b) $f(x) = \begin{cases} a - x & 0 < x < a \\ 0 & a < x \end{cases}$

3. The Fourier transform of $f(x)$ is obtained from Eq. (27) by setting

$K(s,t) = \dfrac{1}{2\pi} e^{-ist}$ and $a = -\infty,\ b = \infty$. Replacing s by λ, show that

Eq. (27) in this case yields $F(\lambda) = \frac{1}{2}(A(\lambda) - iB(\lambda))$, where A and B are given by Eqs. (2) and (3). The inverse transform for this $K(s,t)$ is given by

$f(x) = \displaystyle\int_{-\infty}^{\infty} F(\lambda)e^{i\lambda x} d\lambda.$ Show that this agrees with Eq. (1) of this section.

4. Show that Eq. (18) can be obtained from Eqs. (6) and (17).

5. If $f(x)$ is a continuous differentiable function, find the Fourier coefficients
    of its derivative, $f'(x)$, in terms of $A(\lambda)$ and $B(\lambda)$ as given in Eqs. (2) and (3).
    What are the coefficients of

    $$F(x) = \int_0^x f(t)dt \ ?$$

6. a) Substitute the formulas for $A(\lambda)$ and $B(\lambda)$ into Eq. (1) to show that

    $$f(x) = \frac{1}{\pi} \int_{-\infty}^{\infty} f(t) \int_0^{\infty} \cos \lambda(t - x)d\lambda dt.$$

    (Hint: Change the variable of integration in $A(\lambda)$ and $B(\lambda)$ to t and
    interchange the order of integration.)

    b) Evaluate the inside integral obtained in part a) to show that

    $$f(x) = \lim_{\omega \to \infty} \frac{1}{\pi} \int_{-\infty}^{\infty} f(t) \frac{\sin \omega(t-x)}{t - x} dt.$$

    (Hint: The limit as $\omega \to \infty$ is obtained from the upper limit of the
    integral. Assume the limit can be taken outside of the remaining
    integral.)

    This last notation for $f(x)$ is frequently written as

    $$f(x) = \int_{-\infty}^{\infty} f(t) \delta(t - x)dt$$

    where $\delta(t - x)$ is the Dirac delta function already encountered in
    Section 7.4. The representation of δ as in part b) is an alternate
    to that given previously.

7. Verify that $u(x,t)$ as given in Eq. (24) does satisfy Eq. (21).

8. Find the limit as $t \to \infty$ of the $u(x,t)$ as given in Eq. (24). Notice that this
    value is independent of the original temperature distribution.

9. Solve Eqs. (21), (22) and (23) for each of the following T(x).

a) $T(x) = \begin{cases} 1 & 0 < x < a \\ 0 & a < x \end{cases}$

b) $T(x) = \begin{cases} 1 - x & 0 < x < 1 \\ 0 & 1 < x \end{cases}$

10. Find a formula for the solution of the problem

$$\frac{\partial^2 u}{\partial x^2} = \frac{1}{k} \frac{\partial u}{\partial t} , \qquad 0 < x, \quad 0 < t$$

$$\frac{\partial u}{\partial x} (0,t) = 0, \qquad 0 < t$$

$$u(x,0) = f(x), \qquad 0 < x.$$

11. Find a formula for the solution of the problem

$$\frac{\partial^2 u}{\partial x^2} = \frac{1}{k} \frac{\partial u}{\partial t} , \qquad 0 < t, \quad -\infty < x < \infty$$

$$u(x,0) = f(x) , \qquad -\infty < x < \infty.$$

ANSWERS TO SELECTED EXERCISES

Chapter One

Section 1.1, Page 8

1. a) $y(x) = 2x - 1 + c_1 e^{-x}$

    b) $y(x) = \frac{1}{3} x - \frac{1}{9} + e^{-2x} + c_1 e^{-3x}$

    c) $y(t) = -\frac{1}{2} \sin t - \frac{1}{2} \cos t + c_1 e^{t}$

    d) $y(x) = (x-1)e^{-x} + c_1 e^{-2x}$

2. a) $y(t) = (t-1) + e^{-t}$

    b) $y(t) = 2 - e^{-t/3}$

    c) $i(t) = \frac{E_o}{4L^2 + R^2} (R \sin 2t - 2L \cos 2t) + \frac{2LE_o}{4L^2 + R^2} e^{-Rt/L}$

3. $y_p(t) = 2te^{-t}$

4. a) $y(x) = \frac{c_1}{x^2} - \frac{\cos x}{x^2}$

    b) $y(t) = \frac{c_1}{t} + \frac{\sin t}{t}$

5. $y(t) = 3e^{t^2} - 1$

Section 1.2, Page 13

1. a) $y_h(x) = (c_1 + c_2 x)e^{-2x}$

    b) $y_h(x) = c_1 e^{-3x} + c_2 e^{2x}$

    c) $y_h(x) = e^{x}(c_1 \cos 3x + c_2 \sin 3x)$

    d) $y_h(x) = e^{-x/2}(c_1 \cos \frac{\sqrt{3}}{2} x + c_2 \sin \frac{\sqrt{3}}{2} x)$

2. a) $y(x) = \frac{1}{4} + c_1\cos 2x + c_2\sin 2x$ b) $y(t) = \frac{1}{4} e^{-t} + (c_1 + c_2 t)e^{-3t}$

   c) $y(t) = -\frac{1}{3}\sin 2t + c_1\cos t + c_2\sin t$

3. a) $y = \frac{1}{6} - \frac{5}{3} e^{-3t} + \frac{5}{2} e^{-2t}$ b) $y = e^t - \frac{1}{2} e^{-2t} - t - \frac{1}{2}$

   c) $y = \frac{1}{2}\cos t + \frac{1}{\sqrt{3}}\sin \sqrt{3}\, t - \frac{1}{2}\cos \sqrt{3}\, t$

4. $R = \sqrt{c_1^2 + c_2^2}$, $\phi = \tan^{-1}(c_1/c_2)$

Section 1.3, Page 20

1. $x(t) = (1 - \frac{1}{\omega^2})\cos \omega t + \frac{1}{\omega}\sin \omega t + \frac{1}{\omega^2} = R\sin(\omega t + \phi) + \frac{1}{\omega^2}$

   where $\tan \phi = \dfrac{\omega^2 - 1}{\omega}$ and $R^2 = \dfrac{1}{\omega^2} + \dfrac{(\omega^2 - 1)^2}{\omega^4}$

2. $x(t) = \dfrac{1}{\omega^2 - 4}(\cos 2t - \cos \omega t) = \dfrac{2}{\omega^2 - 4}\sin(\frac{\omega + 2}{2})t\,\sin(\frac{\omega - 2}{2})t$

3. $x(t) = e^{-t}(\frac{1}{5}\cos t + \frac{2}{5}\sin t) - \frac{1}{5}\cos 2t - \frac{1}{10}\sin 2t$

   $= R_1 e^{-t}\sin(t + \phi_1) - R_2\sin(2t + \phi_2)$

   where $\begin{array}{ll} R_1 \cos \phi_1 = 2/5 \\ R_1 \sin \phi_1 = 1/5 \end{array}$ and $\begin{array}{ll} R_2 \cos \phi_2 = 1/10 \\ R_2 \sin \phi_2 = 1/5 \end{array}$

4. a) If $\lambda^2 = n^2\pi^2$, then $y(x) = \sin n\pi x$, $n = 1,2,3, \cdots$.

   b) $y(x) = \dfrac{1 - \cos \lambda x}{\lambda^2} + c_2 \sin \lambda x$, $c_2 = \dfrac{\cos(\lambda)-1}{\lambda^2 \sin \lambda}$, $\lambda \neq (2n - 1)\pi$, $n = 1,2,\cdots$

   If $\lambda = 2n\pi$, $n = 1,2,3,\cdots$, then $y(x)$, as given above, satisfies the
   problem for any c_2.

   c) No solution.

5. a) $y(x) = c_1 \cos \lambda x + c_2 \sin \lambda x + c_3 e^{-\lambda x} + c_4 e^{\lambda x}$

b) $y(t) = (c_1 + c_2 t + c_3 t^2) e^{-t}$

c) $y(t) = c_1 e^{-t} + c_2 e^t + c_3 e^{-2t} - \frac{1}{2} t + \frac{1}{4}$

Section 1.4, Page 27

1. $x_1 = 1.2$, $x_2 = 1.428$, $x_3 = 1.683$, $x_4 = 1.962$, $x_5 = 2.260$, $x_6 = 2.570$

2. Improved Euler $x_1 = 1.214$, $x_2 = 1.457$, $x_3 = 1.727$, $x_4 = 2.019$

    Runge–Kutta $x_1 = 1.215$, $x_2 = 1.459$, $x_3 = 1.730$, $x_4 = 2.023$

3. $y(.3) = 2.289$, $y_3 = 2.278$

4. 1.231 (Euler), 1.271 (Improved Euler), 1.274 (Runge-Kutta)

Chapter Two
Section 2.1, Page 35

1. a) $\begin{bmatrix} -1 & 4 & -2 \\ 3 & -4 & 2 \end{bmatrix}$
   b) $\begin{bmatrix} 8 & 3 & -4 \\ -9 & 7 & -21 \end{bmatrix}$

   c) $\begin{bmatrix} 1 & 0 \\ 3 & -1 \\ -2 & -3 \end{bmatrix}$
   d) $\begin{bmatrix} -3 & -2 & 2 \\ 3 & -2 & 8 \end{bmatrix}$

2. $A = \begin{bmatrix} 1 & 3 & -1 & 1 \\ 2 & -1 & 0 & 1 \\ 5 & 1 & 9 & 0 \end{bmatrix}$, $X = \begin{bmatrix} x_1 \\ x_2 \\ x_3 \\ x_4 \end{bmatrix}$, $B = \begin{bmatrix} 1 \\ 3 \\ 5 \end{bmatrix}$

3. a) $t = 4$ b) 3×5 c) No

4. a) Yes d) No g) No

5. $\begin{bmatrix} 13 & 3 \\ -1 & -4 \\ 5 & -8 \end{bmatrix}$

6. a) $AB = \begin{bmatrix} 18 & 90 \\ 8 & 43 \end{bmatrix}$, $BA = \begin{bmatrix} 34 & 24 & 8 \\ 36 & 27 & 9 \\ 0 & 0 & 0 \end{bmatrix}$ b) $(AB)^T = \begin{bmatrix} 18 & 8 \\ 90 & 43 \end{bmatrix} = B^T A^T$

7. $x = \begin{bmatrix} -6 \\ 17 \\ -8 \end{bmatrix}$

Section 2.2, Page 44

1. -131 4. -6 5. No

6. $\text{adj } A = \begin{bmatrix} -1 & -1 & 1 \\ -1 & 1 & -1 \\ 2 & -2 & 0 \end{bmatrix}$, $A^{-1} = \begin{bmatrix} \frac{1}{2} & \frac{1}{2} & -\frac{1}{2} \\ \frac{1}{2} & -\frac{1}{2} & \frac{1}{2} \\ -1 & 1 & 0 \end{bmatrix}$

7. $\text{adj } A = \begin{bmatrix} -6 & 0 & 0 & 0 \\ 0 & -3 & 0 & 0 \\ 0 & 0 & -2 & 0 \\ 0 & 0 & 0 & 6 \end{bmatrix}$ $A^{-1} = \begin{bmatrix} 1 & 0 & 0 & 0 \\ 0 & 1/2 & 0 & 0 \\ 0 & 0 & 1/3 & 0 \\ 0 & 0 & 0 & -1 \end{bmatrix}$

8. a) 1536 b) 0 c) -200

Section 2.3, Page 51

1. $A^{-1} = \begin{bmatrix} 1/3 & 1/3 \\ -1/9 & 2/9 \end{bmatrix}$

3. a) $A^{-1} = \begin{bmatrix} 37 & 11 & -3 \\ 34 & 10 & -3 \\ 10 & 3 & -1 \end{bmatrix}$ b) $A^{-1} = \frac{1}{2} \begin{bmatrix} 3 & 9 & 5 \\ 2 & 4 & 2 \\ 1 & 1 & 1 \end{bmatrix}$

3. c) $A^{-1} = \begin{bmatrix} 9/10 & -1/2 & -8/5 \\ 7/10 & -1/2 & -4/5 \\ -1/2 & 1/2 & 1 \end{bmatrix}$ d) $A^{-1} = \begin{bmatrix} -23 & 11 & -19/2 & 15/2 \\ -3 & 1 & -1 & 1 \\ 9 & -4 & 4 & -3 \\ -2 & 1 & -1/2 & 1/2 \end{bmatrix}$

4. a) $\begin{bmatrix} 0 & 0 & 1 \\ 0 & 1 & 0 \\ 1 & 0 & 0 \end{bmatrix}$ b) $\begin{bmatrix} 1 & 0 & 0 \\ 0 & 1 & 0 \\ 3 & 0 & 1 \end{bmatrix}$ c) $\begin{bmatrix} 1 & 0 & 0 \\ 0 & 2 & 0 \\ 0 & 0 & 1 \end{bmatrix}$

6. a) 2   b) 2   c) 3   d) 3

7. a) Yes   b) No   c) Yes

Section 2.4, Page 63

1. $\begin{bmatrix} 6 \\ -5 \\ 3 \end{bmatrix}$ 2. $\begin{bmatrix} \frac{1}{2} \\ 0 \\ 0 \end{bmatrix} + c \begin{bmatrix} 1 \\ -1 \\ 1 \end{bmatrix}$

3. No solution.   4. No nontrivial solution.

6. $\begin{bmatrix} 3 \\ -1 \end{bmatrix}$ 7. No solution.

8. $\begin{bmatrix} 5 \\ -3 \\ 2 \\ 0 \end{bmatrix} + c \begin{bmatrix} -3 \\ -3 \\ 4 \\ 2 \end{bmatrix}$ 10. $\begin{bmatrix} 5/3 \\ 2/3 \\ 0 \\ 0 \end{bmatrix} + c_1 \begin{bmatrix} 2 \\ -1 \\ 1 \\ 0 \end{bmatrix} + c_2 \begin{bmatrix} 5 \\ 2 \\ 0 \\ 3 \end{bmatrix}$

Chapter Three

Section 3.1, Page 72

1. a) r, s   b) W   c) $r^2 + 2s^2 \le 2$ d) $0 \le w \le \sqrt{2}$

2. a) $x^2 + y^2 \ne 9$, all z except z = 0   b) all x, y, $0 < z \le 2$

   c) $1 \le x^2 + y^2 < 17$, all z   d) all x, y,   z < 4

3. a) $f_x = -e^{-x} \cos y, \quad f_y = -e^{-x} \sin y$

    b) $f_x = \dfrac{-y}{x^2 + y^2}, \quad f_y = \dfrac{x}{x^2 + y^2}$

    c) $f_x = \dfrac{x}{x^2 + y^2}, \quad f_y = \dfrac{y}{x^2 + y^2}$

4. $f_{xx} = 2\cos y, \; f_{yy} = -x^2 \cos y, \; f_{xy} = f_{yx} = -2x \sin y$

5. a) $4x + 2y - z = 3$ b) $12x + 27y - z = 44$

    c) $4r + w = 4$ d) $4x + y - e^{-2}z = 5$

6. Approximate value: 4.8, actual value: 4.82

8. $f(1.9, 1.2) \approx 5.2, \; f(1.9, 1.4) \approx 5.4, \; f(1.9, .8) \approx 4.5$

Section 3.2, Page 81

1. $\dfrac{\partial z}{\partial t} = -\pi^3/8 \quad \text{at} \quad t = \pi/2$

2. $\dfrac{\partial z}{\partial r} = -\dfrac{4}{3}\cos(4/3), \quad \dfrac{\partial z}{\partial s} = -\dfrac{8}{9}\cos(4/3) \quad \text{at} \quad r = 1, \; s = 0$

3. $\dfrac{dz}{dt} = -3/2 \quad \text{at} \quad t = 0$

4. $\dfrac{dy}{dx} = v\dfrac{du}{dx} + u\dfrac{dv}{dx}$

7. c) $\dfrac{1}{r}\dfrac{\partial}{\partial r}\left(r\dfrac{\partial w}{\partial r}\right) + \dfrac{1}{r^2}\dfrac{\partial^2 w}{\partial \theta^2} = \dfrac{\partial^2 f}{\partial x^2} + \dfrac{\partial^2 f}{\partial y^2}$

8. $J = -\dfrac{1}{3}$ 9. $J = r$

10. a) $\dfrac{\partial y}{\partial u} = -\dfrac{y(y + 1)}{u + uy \ln u}, \quad \dfrac{\partial y}{\partial x} = \dfrac{y}{1 + y \ln u}$

    b) $\dfrac{\partial y}{\partial u} = -\dfrac{xy + 4x^2 u}{xu + \cos y}, \quad \dfrac{\partial y}{\partial x} = \dfrac{yu + 4xu^2}{xu + \cos y}$

11. a) $\frac{\partial u}{\partial y}$ and $\frac{\partial v}{\partial y}$ are given by the equations

$$(2u + 3x)u_y - 2v\,v_y = 2 \quad \text{and} \quad (v - 4)u_y + (u + x\cos v)v_y = -3x$$

Section 3.3, Page 91

1. a) $2x\cos x^4$ b) $-3x^2\ln(3 + x^{3/2})$

   c) $\displaystyle\int_0^{\pi/2x} (\sin xt + x\cos xt)dt - \pi/2x$

   d) $3\alpha^2 e^{-\alpha^7} - 2e^{-4\alpha^3} - \displaystyle\int_{2\alpha}^{\alpha^3} t^2 e^{-\alpha t^2}\,dt$

4. $\tan^{-1}x$

5. a) relative maximum at $(0,0)$ b) saddle point at $(0,0)$

   c) no result at $(0,0)$, saddle point at $(3, -9/2)$

   d) relative minimum at $(0,-1)$

   e) the only critical point is not a relative maximum or relative minimum.

   f) relative maximum at $(0,0,1)$

6. Absolute minimum is 3, no absolute maximum.

7. $\dfrac{1}{\sqrt{3}} \times \dfrac{1}{\sqrt{3}} \times \dfrac{1}{\sqrt{3}}$

8. area = 8

9. a) relative maximum at $(\pm 2,0)$ and relative minimum at $(0,\pm 2)$.

   b) relative maximum at $(0, \sqrt{2}/2, \sqrt{2}/2)$ and relative minimum at $(0, -\sqrt{2}/2, -\sqrt{2}/2)$

   c) relative maximum at $(\pm\sqrt{2/3}, \pm\sqrt{2/3}, \sqrt{1/3})$ and relative minimum at $(0,0, \pm 1)$

10. f(x,y,z) has a maximum of 108 and a minimum of 0.

11. Hint: See Example 9. and Exercise 9c.

Chapter 4

Section 4.1, Page 101

1. a) 5/6   b) $\pi/4$ c) $\pi^2/2 + 2$ d) 4/3

2. a) $\displaystyle\int_{-2}^{2} \int_{0}^{\sqrt{2-x^2/2}} y\, dy\, dx$ b) $\displaystyle\int_{1}^{e^3} \int_{\ln y}^{3} xy\, dx\, dy$

d) $\displaystyle\int_{-3}^{0} \int_{-\sqrt{1-x}}^{x+1} e^{-x^2}\, dy\, dx \;+\; \int_{0}^{1} \int_{-\sqrt{1-x}}^{+\sqrt{1-x}} e^{-x^2}\, dy\, dx$

3. a) 1/3   b) 2/15

4. $A = 2\displaystyle\int_{0}^{a} \int_{\sqrt{a^2 - x^2}}^{a} dy\, dx = 2a^2 - \pi a^2/2$

5. $V = \displaystyle\int_{0}^{1} \int_{0}^{2-2x} (2 - 2x - y)\, dy\, dx = \dfrac{2}{3}$

6. $2a^3/3$

7. a) 1/4   b) $\cos t - \cos(2 - t)$ c) π

Section 4.2, Page 110

1. a) 26   b) 15/64   2. a) 6   b) 4/3

3. a) $\displaystyle\int_{0}^{1} \int_{0}^{2\sqrt{1-y^2}} \int_{0}^{x+2} dz\, dx\, dy$ b) $\displaystyle\int_{0}^{2} \int_{0}^{x+2} \int_{0}^{\sqrt{4-x^2/2}} dy\, dz\, dx$

c) $\displaystyle\int_0^2 \int_0^2 \int_0^{2\sqrt{1-y^2}} dx\,dy\,dz \;+\; \int_2^4 \int_0^{\sqrt{4-(z-2)^2/2}} \int_{z-2}^{2\sqrt{1-y^2}} dx\,dy\,dz$

4. $\sqrt{\pi}/2$

5. $\dfrac{4}{3}\pi\,(8 - 3^{3/2})$

6. $27\pi(2^{3/2} - 1)/2$

7. $248\pi/15$

8. $k\,a^4\pi$

9. 2

10. $(1 - \cos 1)/2$

Chapter Five

Section 5.1, Page 118

3. Approximately 24 miles south-southeast

4. Yes, yes

6. a) $\sqrt{21}$ b) $2/\sqrt{21}$, $-4/\sqrt{21}$, $1/\sqrt{21}$ d) $(2i - 4j + k)/\sqrt{21}$

7. a) $\sqrt{61}$ b) 0, $5/\sqrt{61}$, $6/\sqrt{61}$ d) $(5j + 6k)/\sqrt{61}$

8. a) $\gamma = 90°$ b) $\alpha = 60°$, $150°$

9. a) No   b) No   c) 3.0, 4.5   d) 2   e) infinite   f) yes

Section 5.2, Page 125

1. a) $4j + 3k$ b) -2 c) 5   d) $\sqrt{34}$

   e) $s = \pm 1/3$ f) $7\sqrt{11}/33$ g) $7\sqrt{11}/11$

2. a) $R = (i + 3j + k) + t(3i + 2j - 2k)$

   c) $R = (i + 2j + 5k) + t(i + j + 3k)$

   d) $R = (-i + 3j + 2k) + t(i - j + 2k)$ and $x = -1 + t$, $y = 3 - 5$, $z = 2 + 2t$

4. 3 units of work

5. a) $2x - y + 3z = 6$ b) $x + 2y - 3z = 0$ c) $x - 3y - 2z = 11$

6. $(2i + j + k)/\sqrt{6}$ 7. No

8. A circle and a sphere   10. 5

11. $2\sqrt{14}/7$

Section 5.3, Page 131

1. a) -31 b) $9i - 4j - k$ c) $6i - j - 4k$

    d) $5i - 4j - k$ e) $(i + 3j - 7k)/\sqrt{59}$

2. $4x - 3y - 2z = 0$

3. $R = (i + 2j - k) + t(i + 10j + 3k)$

4. a) $\sqrt{101}$ b) $8x + y + 6z = 37$

7. 0

8. $20\sqrt{5}/3$

9. $(A \cdot C \times D)B - (B \cdot C \times D)A$

Section 5.4, Page 140

1. a) $|R'| = |-a\omega \sin(\omega t)i + a\omega \cos(\omega t)j| = |a\omega|, \quad R'' = -\omega^2 R$

    c) R' is perpendicular to R, R'' is perpendicular to R' in opposite
        direction to R

    d) $R = a \cos(s/a)i + a \sin(s/a)j$

2. $R \times R''$ 3. $t = 0$

4. b) $T = ((\cos t - \sin t)i + (\sin t + \cos t)j)/\sqrt{2}$

    d) Replace t by $\ln(1 + s/\sqrt{2})$ in the given $R(t)$

6. a) $R' = 3i + 4tj$, $R'' = 4j$

    b) $T = (3i + 4tj)/\sqrt{9 + 16t^2}$

    c) $T' = (-48ti + 36j)/(9 + 16t^2)^{3/2}$

    d) $R'' = sT + mT'$ where $s = 16t/\sqrt{9 + 16t^2}$ and $m = \sqrt{9 + 16t^2}$

9. elliptical helix with $z = e^t$

Section 5.5, Page 146

1. $v = 3\pi u_\theta$, $a = -3\pi^2 u_r$

2. $v = 8t u_\theta$, $a = -16t^2 u_r + 8u_\theta$, centripetal acceleration is $-16t^2$ and
    the transverse acceleration is 8.

3. $v = -b \sin(t)u_r + b(1 + \cos t)(-e^{-t})u_\theta$

    $a = (-b \cos t - be^{-2t}(1 + \cos t))u_r + (be^{-t}(1 + \cos t) + 2be^{-t} \sin t)u_\theta$

4. 4   5. $-10/3$ 6. $-11/\sqrt{2}$, $\pm \sqrt{209}$ 9. $\sqrt{2}$

10. a) $\nabla f = (y \cos xy + ze^x)i + x \cos(xy)j + e^x k$

    c) $\nabla f = (2ye^{yz} + 2x \ln y)i + (2xe^{yz} + 2xyze^{yz} + x^2/y)j + 2xy^2 e^{yz} k$

    d) $\nabla f = \dfrac{dg}{dr} \nabla r = \dfrac{dg}{dr} \dfrac{R}{|R|}$, note $R = xi + yj + zk$

11. $f = (x + 1)z + y^2 + c$

Section 5.6, Page 154

1. $i + 8j + 4k$ 2. $\sqrt{(e + \pi)^2 + 1}$, $(\pi + e)i + j$

3. 6   4. $\nabla f = -4i - 2j$, $u = (-i + 2j)/\sqrt{5}$

5. a) $\pm(i - j)$ b) $-i - j$ c) $i + j$

6. b) $(2i + j + 7k)/3\sqrt{6}$ c) $(i + j - \sqrt{2}k)/2$

7. $\theta = \cos^{-1}(13/3\sqrt{21})$ = angle between line and normal to ellipsoid

10. a) $y(2x - 1)$, $2xzi - 2yzj - xk$ b) 3, 0

    c) $3y^2 + \cos z$, 0

    d) $y \cos xy + xe^{xy} - 2yz \sin(z^2)$, $\cos z^2 i + (ye^{xy} - x \cos xy)k$

11. a) $2y^3 z^{-1} + 6x^2 yz^{-1} + 2x^2 y^3 z^{-3}$

    b) $f_{xx} + f_{yy} + f_{zz}$

Section 5.7, Page 166

1. a) 43/6   b) 37/6   c) 28/3

    d) 4π e) -4π f) $8\pi^3/3 - 5\pi$

6. a) 3   b) $264 - 1/5 + \pi^3/24$

8. a) 70π b) 1/20   c) $-\pi$

10. a) -2π b) 0

Section 5.8, Page 178

3. a) $8\pi \sqrt{2}$ b) 1   c) 7/3   d) 0

5. $(\cos(\theta)i + \sin(\theta)j)ad\theta\ dz$

6. $(\sin(\phi)\cos(\theta)i + \sin(\phi)\sin(\theta)j + \cos(\phi)k)\ a^2\sin(\phi)d\phi d\theta$

7. 0   8. 9π

Section 5.9, Page 186

3. a) 42   b) 89/40   c) 2π

5. a) 0   b) 0   c) 0

6. a) π b) 0

Chapter Six

Section 6.1, Page 198

1. a) $-4 + i$ b) $-19 - 17i$ c) $-23/50 - 11i/50$

   d) $\sqrt{13}$, $\sqrt{50}$ e) $-24 - 5i$ f) $-19 + 17i$

2. a) $2 + 4i$ b) $\frac{1}{2} + \frac{1}{2}i$ c) $-4/5 - 8i$ d) $3i$

4. a) on a circle of radius 2, center at $x = 1$, $y = 0$

   b) on the vertical line $x = -2$ c) on the horizontal line $y = 3$

   d) on the vertical line $x = 1$ e) on the ray $\theta = 60°$

   f) on the parabola $y^2 = 4 - 2x$

5. a) $2(\cos 60° - i \sin 60°)$, $2e^{-\pi i/3}$ b) $\sqrt{2}(\cos 135° + i \sin 135°)$, $\sqrt{2}\, e^{3\pi i/4}$

   c) $2(\cos 0° + i \sin 0°)$, $2e^{0i}$ d) $5(\cos 233.1° + i \sin 233.1°)$, $5e^{1.295i}$

6. a) $\frac{\sqrt{2}}{2}\, e^{7\pi i/12}$,   b) $-\frac{1}{2} - i\frac{\sqrt{3}}{2}$

   c) $2^{1/6}\, e^{\pi i/12}$, $2^{1/6}\, e^{3\pi i/4}$, $2^{1/6}\, e^{17\pi i/12}$ e) $\pm(\frac{\sqrt{2}}{2} + \frac{\sqrt{2}}{2}i)$

Section 6.2, Page 205

1. a) $2i - 2$, $1 + 4i$ b) $u = -2$, $v = 2x^2$

2. a) Re $w = x^2 - y^2 + x - y$, Im $w = 2xy + y$

    d) Re $w = \dfrac{2x + x^2 + y^2}{(2+x)^2 + y^2}$,   Im $w = \dfrac{2y}{(2+x)^2 + y^2}$

3. a) $|\arg w| \le \pi/3$ c) $\frac{1}{2} < |w| < 1$ d) $|w - 1| \le 4$

4. a) if $z + 1 = r_1 e^{i\theta_1}$ and $z - 1 = r_2 e^{i\theta_2}$ then $f(z) = (r_1 r_2)^{\frac{1}{2}} e^{i(\theta_1 + \theta_2)/2}$

5. a) $1 - 24i$ b) 0 c) No Limit   d) 2 h) 0

6. a) $z = \pm 2i$ c) No for part a), Yes for part b)

7. a) $6z + 2$ b) $\dfrac{4z}{\sqrt{1 + 4z^2}}$ d) $\dfrac{-2}{(z - 1)^2}$

    c) $4z(z^2 - 1)(z^3 - z + 1)^3 + 3(z^2 - 1)^2(z^3 - z + 1)^2(3z^2 - 1)$

Section 6.3, Page 217

1. a) all z b) all $z \neq 0$ c) No z

    f) w analytic for $x^2 > y^2$ and $xy > 0$ or for $x^2 < y^2$ and $xy < 0$

3. $v(x,y) = \dfrac{x}{x^2 + y^2} + C$

6. a) for $z = 1 + i$ $|f(z)| = 1$, $\arg f(z) = \tan^{-1}(1/3) - \tan^{-1}(2) - \pi/4$

        for $z = -1 - i$ $|f(z)| = 1/\sqrt{5}$, $\arg f(z) = \pi/2 - \tan^{-1}2$

        for $z = -1$ $|f(z)| = \frac{1}{2}$, $\arg f(z) = \pi$

    b) $z = 0$, $\pm i$

8. b) $z = \ln 3 + i\pi$ c) $(1 + i)(\sqrt{2}/2)e^{-3}$

    d) $e^{x^2 - y^2}$ e) Re $z > 0$

9. a) $\ln \sqrt{2} - i\pi/4$ c) $(1 + i\sqrt{3})/2$

16. $w = i/z$

17. b) $1 < \rho < e$, $-\pi < \phi < \pi$ or $1 < |w| < e$ except for negative real axis

Section 6.4, Page 227

2. $\displaystyle\sum_{n=1}^{\infty} \frac{(-1)^{n+1}(z-1)^n}{n}$, $|z - 1| < 1$ 3. $\displaystyle\sum_{n=0}^{\infty} \frac{(-1)^n z^{n+1}}{(n+1)}$

4. a) $\displaystyle\sum_{n=0}^{\infty} (-1)^n \frac{z^n}{4^{n+1}}$ b) $\displaystyle\sum_{n=0}^{\infty} (-1)^n \frac{4^n}{z^{n+1}}$ c) $\displaystyle\sum_{n=0}^{\infty} (-1)^n \frac{z^{n+1}}{4^{n+1}}$

5. $-\displaystyle\sum_{n=0}^{\infty} z^{2(n-1)} = -\frac{1}{z^2} - 1 - z^2 - z^4 - \cdots$, so at $z = 0$ there is a pole of

order two.

6. a) $|z - 2| < 2$ c) all z d) $|z - i/2| < \frac{1}{2}$

9. a) $-\displaystyle\sum_{n=0}^{\infty} \frac{z^{n-3}}{3^{n+1}} = -\frac{1}{3z^3} - \frac{1}{9z^2} - \cdots , 0 < |z| < 3$, $z = 0$ is a third order pole.

d) $\displaystyle\sum_{n=0}^{\infty} \frac{z^{1-2n}}{n!} = z + \frac{1}{z} + \frac{1}{2z^3} + \cdots$, $0 < |z|$, $z = 0$ is an essential singularity.

10. a) $\dfrac{z^2}{3}\left(\displaystyle\sum_{n=0}^{\infty} \frac{(-1)^n z^n}{2^{n+1}} + \sum_{n=0}^{\infty} z^n\right) = \frac{z^2}{2} + \frac{z^3}{4} + \frac{3z^4}{8} + \cdots$

b) $\dfrac{z^2}{3}\left(\displaystyle\sum_{n=0}^{\infty} \frac{(-1)^n z^n}{2^{n+1}} - \sum_{n=0}^{\infty} \frac{1}{z^{n+1}}\right) = \cdots - \frac{1}{3z^2} - \frac{1}{3z} - \frac{1}{3} - \frac{z}{3} + \frac{z^2}{6} - \frac{z^3}{12} \cdots$

c) $\dfrac{z^2}{3}\left(\dfrac{1}{z}\displaystyle\sum_{n=0}^{\infty} \frac{(-1)^n 2^n}{z^n} - \sum_{n=0}^{\infty} \frac{1}{z^{n+1}}\right) = -1 + \frac{1}{z} - \frac{3}{z^2} + \frac{5}{z^3} - \cdots$

Section 6.5, Page 236

1. $2/3 + 11i/3$

2. $88/15 + 2i/3$, $20/3 + 14i/3$

3. $-1 + i$, 0

4. 0

5. a) $20/3 + 5i/3$

b) $\frac{1}{2} \ln 5 + (\pi - \tan^{-1} 2)i$

c) $1 - \sin 1 + ie(\cos 1)$

d) $\frac{1}{2} \ln .6$

6. $8\pi i$

7. a) 0   b) $2\pi i$ c) 0

8. f) $-\pi i$, πi g) 0

9. a) $\dfrac{w}{w^2 + 2w + 3}$

b) $2w^3 e^{w^2} - 9we^{-3w}$

c) $\displaystyle\int_1^{2w} \dfrac{dz}{(z-w)^2} + \dfrac{2}{w}$

d) $\displaystyle\int_c \dfrac{zdz}{(z-w)^2}$

Section 6.6, Page 244

1. a) $2\pi i$, $6\pi i$ b) 0

2. a) $2\pi i$ b) πi c) $-2\pi i$ d) $-\pi i/4$

e) $2\pi i \left(-\dfrac{1}{25} + \dfrac{i-1}{(2-i)^2(-2i)} + \dfrac{-i-1}{(2+i)^2(2i)} \right) = 0$

3. a) $m = 2$, residue $= -2$ at $z = 0$ c) $m = 2$, residue $= e$ at $z = 1$

b) $m = 2$, residue $= -1/9$ at $z = 0$ d) $m = 2$, residue $= -1/4$ at $z = 1$

   $m = 1$, residue $= 1/9$ at $z = -3$ $m = 2$, residue $= 1/4$ at $z = -1$

e) $m = 1$, residue $= 3/2$ at $z = 0$ f) $m = 1$, residue $= -1/4$ at $z = -i$

   $m = 1$, residue $= 1$ at $z = 1$ $m = 1$, residue $= -1/4$ at $z = i$

   $m = 1$, residue $= -5/2$ at $z = 2$ $m = 1$, residue $= 1/4$ at $z = -1$

                                              $m = 1$, residue $= 1/4$ at $z = 1$

4. a) $\dfrac{1}{z} + \dfrac{5/3}{z+4} - \dfrac{5/3}{z-1}$

b) $\dfrac{1}{z} - \dfrac{1}{z^2} + \dfrac{1}{z^3} - \dfrac{1}{z+1}$

5. $B = -1$ $C = -1$

Chapter 7

Section 7.1, Page 251

3. a) $\dfrac{6}{s + 2}$

    b) $\dfrac{5}{s^2} + \dfrac{2s}{s^2 + 9}$

    c) $(2s + 6)Y(s) - y(0)$

    e) $(s^2 + 2s + 1)Y(s) - sy(0) - y'(0)$

4. a) $\dfrac{2}{s^3}$

    b) $\dfrac{1}{(s - a)^2}$

5. a) $2 - 3t$

    b) $-3 + \dfrac{5}{2} \sin 2t$

    c) $1 - e^{-t}$

Section 7.2, Page 257

1. a) $y(t) = \dfrac{1}{9} e^{-3t} + \dfrac{1}{3} t + \dfrac{8}{9}$

    b) $\overset{\circ}{y}(t) = \dfrac{2}{5} e^{-t} + \dfrac{1}{5} \sin 2t - \dfrac{2}{5} \cos 2t$

2. b) $y(t) = \cos t + \dfrac{2}{3} \sin t - \dfrac{1}{3} \sin 2t$ c) $y(t) = 2 - 2(t + 1)e^{-t}$

3. a) $x(t) = \dfrac{3}{2} e^{-2t} + \dfrac{2}{9} e^{-3t} + \dfrac{2}{3} t - \dfrac{13}{18}$, $y(t) = \dfrac{3}{4} e^{-2t} + \dfrac{2}{9} e^{-3t} + \dfrac{1}{6} t + \dfrac{1}{36}$

    b) $x(t) = -2 + 2 \cos t$, $y(t) = t - \sin t$

Section 7.3, Page 265

1. c) $\dfrac{12}{(s + 3)^5}$

    d) $\dfrac{s - a}{s^2 - 2as + a^2 + \omega^2}$

2. b) $\dfrac{s^2 - \omega^2}{(s^2 + \omega^2)^2}$

    d) $\dfrac{6}{(s + 2)^4}$

4. a) $\dfrac{1}{3} e^{-t/3} \cos(2t/3)$

    b) $\dfrac{3}{10} e^{-2t/5} (t/5)^2$

5. a) $y(0) = 0$, $y(\infty) = 1$

    b) $y(0) = 0$, $y(\infty) = 1/6$

    c) $y(0) = 2$, $y(\infty) = 1$

    d) $y(0) = 0$, $y(\infty)$ does not exist

6. a) $y(t) = \frac{1}{4} - \frac{1}{4} e^{-2t}(\cos 2t + \sin 2t)$

    b) $y(t) = \cos t + 2 \sin t - 2e^{-t} \sin t$

    c) $y(t) = \frac{1}{\omega^2 - 9} (\frac{\omega}{3} \sin 3t - \sin \omega t), \quad \omega^2 \neq 9,$

        $y(t) = \frac{1}{18} \sin 3t - \frac{1}{6} t \cos 3t, \quad \omega = 3$

    d) $y(t) = 4 - 2(t + 1)e^{-t}$

Section 7.4, Page 272

1. a) $f(t) = 2u(t,1) - 3u(t,3), \quad F(s) = \frac{2e^{-s} - 3e^{-3s}}{s}$

    c) $f(t) = 1 + u(t,1)((t - 1)^2 + 2(t - 1)), \quad F(s) = \frac{1}{s} + \frac{2(s + 1)}{s^3} e^{-s}$

    d) $f(t) = u(t,\pi/2)\cos(t - \pi/2), \quad F(s) = \frac{se^{-\pi s/2}}{s^2 + 1}$

    e) $f(t) = e^{-1/3} u(t,1)e^{-(t - 1)/3}, \quad F(s) = \frac{e^{-(s + 1/3)}}{s + 1/3}$

2. a) $1 = u(t,2)$ b) $u(t,1)e^{-3(t - 1)}$

    c) $\frac{1}{2} u(t,\pi)\sin 2(t - \pi)$ d) $\frac{1}{2} u(t,1)(1 - e^{-2(t - 1)})$

3. a) $y(t) = \begin{cases} -\frac{1}{2} - \frac{1}{2}e^{-t}(\cos t + \sin t) & 0 \le t < 1 \\ - \frac{1}{2}e^{-(t-1)}(\cos(t - 1) + \sin(t - 1)) - \frac{1}{2}e^{-t}(\cos t + \sin t) & 1 \le t \end{cases}$

    b) $y(t) = \begin{cases} \cos t & 0 \le t < \pi/2 \\ \cos t + 1 - \sin t & \pi/2 \le t < \pi \\ - \sin t & \pi \le t \end{cases}$

    d) $y(t) = \begin{cases} t & 0 \le t < 1 \\ 1 + \sin(t - 1) & 1 \le t \end{cases}$

4. b) $y(t) = e^{-(t - \pi)} \sin(t - \pi),$ $t \geq \pi$

    c) $y(t) = 2e^{-t} - e^{-2t} + u(t,2) \; (e^{-(t - 2)} - e^{-2(t - 2)})$

Section 7.5, Page 282

1. a) $f(t) = \frac{1}{3} - \frac{1}{3} e^{-3t}$ b) $f(t) = \frac{1}{8} t^2 - \frac{1}{16} + \frac{1}{16} \cos 2t$

    c) $f(t) = -\frac{2}{5} e^{-2t} + \frac{2}{5} \cos t + \frac{1}{5} \sin t$ d) $f(t) = 1 - e^{-t} \cos t$

2. a) $F(s) = \dfrac{s}{(s + 2)(s^2 + 9)}$ b) $F(s) = \dfrac{2}{s^3(s^2 + 1)}$

3. $y(t) = \frac{1}{9} t - \frac{1}{27} \sin 3t$

4. $y(t) = \displaystyle\int_0^t e^{-(t - u)} \sin(t - u) \; f(u)du$

5. $Y(s) = \dfrac{F(s)}{1 + K(s)}$

6. a) $y(t) = \sin t$ b) $y(t) = -\frac{2}{7} \cos\sqrt{2} \; t + \frac{9}{7} \cos 3t$

    c) $y(t) = \frac{2}{3} + \frac{1}{3} \cos\sqrt{6} \; t$

7. a) $f(t) = 1 - e^{-t}$ b) $5t^2 e^{-3t}$

    c) $f(t) = \frac{1}{5} e^{-t} + \dfrac{e^{-2it}}{-8 - 4i} + \dfrac{e^{2it}}{-8 + 4i} = \frac{1}{5} e^{-t} - \frac{1}{5} \cos 2t + \frac{1}{10} \sin 2t$

Chapter 8

Section 8.1, Page 293

3. a) $P = 2/3$ c) $P = 3$

    d) $P = \pi/3$ f) $P = 2\pi$

9. a) $f(x) = \dfrac{4}{\pi} \displaystyle\sum_{n=1}^{\infty} \dfrac{\sin(2n - 1)x}{2n - 1}$

b) $f(x) = \dfrac{\pi}{2} - \dfrac{4}{\pi} \sum\limits_{n=1}^{\infty} \dfrac{\cos(2n-1)x}{(2n-1)^2}$

c) $f(x) = \dfrac{1}{2} + \dfrac{2}{\pi} \sum\limits_{n=1}^{\infty} \dfrac{\sin(2n-1)x}{2n-1}$

d) $f(x) = \dfrac{\pi^2}{6} + 2 \sum\limits_{n=1}^{\infty} \dfrac{(-1)^n \cos nx}{n^2} + \sum\limits_{n=1}^{\infty} \left(\dfrac{2}{n^3\pi}(\cos n\pi - 1) - \dfrac{\pi}{n}(-1)^n\right)\sin nx$

e) $f(x) = \sin 3x$

Section 8.2, Page 304

1. a) even   b) even   c) neither
    d) even   e) even   f) odd

2. a) $f(x) = -\dfrac{4}{\pi^2} \sum\limits_{n=1}^{\infty} \dfrac{\cos\dfrac{(2n-1)\pi x}{2}}{(2n-1)^2} + \sum\limits_{n=1}^{\infty} \left(\dfrac{1-3\cos n\pi}{n\pi}\right)\sin\dfrac{n\pi x}{2}$

    b) $f(x) = \dfrac{1}{3} + \dfrac{4}{\pi^2} \sum\limits_{n=1}^{\infty} \dfrac{(-1)^n \cos n\pi x}{n^2}$

    c) $f(x) = \dfrac{6}{\pi} \sum\limits_{n=1}^{\infty} \dfrac{\sin\dfrac{n\pi x}{3}}{n}$

    d) $f(x) = \dfrac{e-e^{-1}}{2} + (e-e^{-1}) \sum\limits_{n=1}^{\infty} \dfrac{(-1)^n \cos n\pi x - n\pi(-1)^n \sin n\pi x}{1+n^2\pi^2}$

    e) $f(x) = 1$ f) $f(x) = \sin \pi x$

7. $f(x) = \dfrac{3}{2} - \dfrac{4}{\pi^2} \sum\limits_{n=1}^{\infty} \dfrac{\cos(2n-1)\pi x}{(2n-1)^2}$, $f(x) = 2 \sum\limits_{n=1}^{\infty} \dfrac{(1-2\cos n\pi)}{n\pi} \sin n\pi x$

8. $f(x) = \dfrac{\ell}{2} + \dfrac{4\ell}{\pi^2} \sum\limits_{n=1}^{\infty} \dfrac{\cos\dfrac{(2n-1)\pi x}{\ell}}{(2n-1)^2}$, $f(x) = \dfrac{2\ell}{\pi} \sum\limits_{n=1}^{\infty} \dfrac{\sin\dfrac{n\pi x}{\ell}}{n}$

9. $f(x) = \frac{1}{2} - \frac{1}{\pi} \sum_{n=1}^{\infty} \frac{\sin 2n\pi x}{n}$

10. Yes, not necessarily, check at $x = 0$ and $x = a$.

Section 8.3, Page 314

1. b) $f(x)$ is piecewise smooth and continuous for $-1 < x < 1$, therefore its Fourier series converges to the stated function for $-1 < x < 1$ and to $\frac{1}{2}$ for $x = \pm 1, \pm 3, \cdots$.

   c) $f(x)$ is piecewise smooth and continuous for $-1 < x < 1$, therefore its Fourier series converges to the stated function for $-1 < x < 1$ and to $\frac{1}{2}(\sin 1 + \sin -1) = 0$ for $x = \pm 1, \pm 3, \cdots$.

2. The cosine series converges to 1, 3/4, 3/4 at $x = 0, 1, -1$ respectively. The sine series converges to 0, 0, 0 at $x = 0, 1, -1$ respectively.

5. a) $y(x) = \frac{54}{\pi} \sum_{n=1}^{\infty} \frac{(-1)^{n+1} \sin \frac{n\pi x}{3}}{n\pi(18 - n^2\pi^2)}$

   b) $y(x) = 2 \sum_{n=1}^{\infty} \frac{n(1 - (-1)^n e^{\pi})}{(1 + n^2)(1 - 2n^2)} \sin nx$

   c) $y(x) = \frac{2}{3 - 9\pi^2} \sin 3\pi x$

6. $A_0 = \frac{a_0}{\beta}$, $A_n = \dfrac{(\beta - \frac{n^2\pi^2}{a^2})a_n - \frac{n\pi\alpha}{a} b_n}{(\beta - \frac{n^2\pi^2}{a^2})^2 + \frac{n^2\pi^2\alpha^2}{a^2}}$, $B_n = \dfrac{\frac{n\pi\alpha}{a} a_n + (\beta - \frac{n^2\pi^2}{a^2})b_n}{(\beta - \frac{n^2\pi^2}{a^2})^2 + \frac{n^2\pi^2\alpha^2}{a^2}}$

7. $y_p(t) = \frac{1}{\beta} + \frac{2}{\pi} \sum_{n=1}^{\infty} \dfrac{(2n-1)\pi\alpha \cos(2n-1)\pi t + (\beta - (2n-1)^2\pi^2)\sin(2n-1)\pi t}{(2n-1)((\beta - (2n-1)^2\pi^2)^2 + (2n-1)^2\pi^2\alpha)}$

8. a) $u(x,t) = \frac{3}{2} + \frac{2}{\pi} \sum_{n=1}^{\infty} \frac{(-1)^n}{(2n-1)} e^{-\beta_n t} \cos \frac{n\pi x}{\ell}$

b) $u(x,t) = \dfrac{\ell}{2} - \dfrac{4\ell}{\pi^2} \displaystyle\sum_{n=1}^{\infty} \dfrac{1}{(2n-1)^2} e^{-\beta_{2n-1}t} \cos \dfrac{(2n-1)\pi x}{\ell}$

10. $u(x,t) = \displaystyle\sum_{n=1}^{\infty} c_n e^{-\beta_n t} \sin \dfrac{n\pi x}{\ell}$, $c_n = \dfrac{2}{\ell} \displaystyle\int_0^{\ell} T(x) \sin \dfrac{n\pi x}{\ell} dx$

11. $u(x,t) = \dfrac{a_0}{2} + \displaystyle\sum_{n=1}^{\infty} (a_n \cos \dfrac{n\pi c}{\ell} t + b_n \sin \dfrac{n\pi c}{\ell} t) \sin \dfrac{n\pi x}{\ell}$,

where $a_n = \dfrac{2}{\ell} \displaystyle\int_0^{\ell} f(x) \sin \dfrac{n\pi x}{\ell} dx$ and $b_n = \dfrac{2}{n\pi c} \displaystyle\int_0^{\ell} g(x) \sin \dfrac{n\pi x}{\ell} dx$

Section 8.4, Page 324

1. a) $f(x) = \displaystyle\int_0^{\infty} \{\dfrac{\sin \lambda a}{2\pi\lambda} \cos \lambda x + \dfrac{(1-\cos \lambda a)}{2\pi\lambda} \sin \lambda x\} d\lambda$

b) $f(x) = \dfrac{2}{\pi} \displaystyle\int_0^{\infty} \dfrac{\cos \lambda x}{1 + \lambda^2} d\lambda$

c) $f(x) = \displaystyle\int_0^{\infty} e^{-\lambda} \sin(\lambda x) d\lambda$

d) $f(x) = (\sqrt{2/\pi}) a \displaystyle\int_0^{\infty} e^{-a^2\lambda^2/2} \cos(\lambda x) d\lambda$

2. a) $f(x) = \dfrac{2}{\pi a} \displaystyle\int_0^{\infty} \dfrac{\sin \lambda a}{\lambda} \cos(\lambda x) d\lambda$, $f(x) = \dfrac{2}{\pi a} \displaystyle\int_0^{\infty} \dfrac{(1-\cos \lambda a)}{\lambda} \sin(\lambda x) d\lambda$

b) $f(x) = \dfrac{2}{\pi} \displaystyle\int_0^{\infty} \dfrac{(1-\cos \lambda a)}{\lambda^2} \cos(\lambda x) d\lambda$, $f(x) = \dfrac{2}{\pi} \displaystyle\int_0^{\infty} \dfrac{a\lambda - \sin a\lambda}{\lambda^2} \sin(\lambda x) d\lambda$

5. For $f'(x)$ the cosine coefficient is $\lambda B(\lambda)$ and the sine coefficient is $-\lambda A(\lambda)$.
    For $F(x)$ the cosine coefficient is $-B(\lambda)/\lambda$ and the sine coefficient is $A(\lambda)/\lambda$,
    if $F(x) \to 0$ as $x \to \infty$.

8. $\displaystyle\lim_{t \to \infty} u(x,t) = 0$

9. a) $u(x,t) = \dfrac{2}{\pi} \displaystyle\int_0^\infty \dfrac{1 - \cos \lambda a}{\lambda} \sin(\lambda x) e^{-\lambda^2 kt} \, d\lambda$

    b) $u(x,t) = \dfrac{2}{\pi} \displaystyle\int_0^\infty \dfrac{\lambda \cos \lambda - \sin \lambda}{\lambda^2} \sin(\lambda x) e^{-\lambda^2 kt} \, d\lambda$

10. $u(x,t) = \displaystyle\int_0^\infty A(\lambda) \cos(\lambda x) e^{-\lambda^2 kt} \, d\lambda$ where $A(\lambda) = \dfrac{2}{\pi} \displaystyle\int_0^\infty f(x) \cos(\lambda x) dx$

11. $u(x,t) = \displaystyle\int_0^\infty \{A(\lambda) \cos \lambda x + B(\lambda) \sin \lambda x\} e^{-\lambda^2 kt} \, d\lambda$

    where $A(\lambda)$ and $B(\lambda)$ are given by Eqs. (2) and (3).

INDEX